Total Quality Through Project Management

Total Quality Through Project Management

Jeffrey S. Leavitt, CQE, PMP

Philip C. Nunn, PMP

McGraw-Hill, Inc.

New York San Francisco Washington, D.C. Auckland Bogotá
Caracas Lisbon London Madrid Mexico City Milan
Montreal New Delhi San Juan Singapore
Sydney Tokyo Toronto

Library of Congress Cataloging-in-Publication Data

Leavitt, Jeffrey S.
Total quality through project management / Jeffrey S. Leavitt, Philip C. Nunn.
p. cm.
Includes index.
ISBN 0-07-036980-1
1. Industrial project management. 2. Total quality management. I. Nunn, Philip C. II. Title.
HD69.P75L4 1994
658.5'62—dc20 93-38403
CIP

1 2 3 4 5 6 7 8 9 0 DOC/DOC 9 9 8 7 6 5 4

ISBN 0-07-036980-1

The sponsoring editor for this book was Larry Hager and the production supervisor was Pamela A. Pelton. This book was set in Century Schoolbook. It was composed by North Market Street Graphics.

Printed and bound by R. R. Donnelley & Sons Company.

To Nathaniel Scott and Erin Marie:
May you always find the peace, joy, and happiness
you both deserve throughout "Project Life."

Jeffrey S. Leavitt

Contents

Preface

One of the most commonly used words in business today is "quality." You see it in advertisements and you hear it from company executives. There have been hundreds of books published on the topic of quality, some from the world's leading authorities on the subject. So why do we need another book on quality? That is exactly the point; we probably do not need "just another book" on quality. *Total Quality Through Project Management* is unique in several respects. First, the book focuses on the use of project management to implement Total Quality Management (TQM). Project management has been beneficial in many areas, but has had limited exposure in the field of quality to date. The reason for this sporadic use is that the field of project management is itself in an evolutionary growth phase. Many uses of project management have yet to be discovered. Second, the material is presented in a very structured manner. A description of project management is provided, along with the eight key steps through which every project goes.

Although Phil Nunn wrote the project management part of this book in a year, it took him 12 years to develop the concept that the implementation of project management was a systematic process. While teaching project management during those years, he continued to refine an approach to project management which communicated the concepts and principles more clearly. This book contains the result of that prolonged search for a better way. It envelops the concepts, principles, and practices of project management in a process. This process guides a person through eight steps to plan and manage a project. If followed, this process will help implement a Total Quality Management program successfully, as well as any other industrial project. After using this process several times, project managers will recognize it as a beginning, not an ending, and will develop variations which are tuned to their company's culture.

The book is designed for the quality practitioner or individual who has been charged with the TQM responsibility. Referring to a book on project management might provide some insight; however, the direct application to quality would be lacking. This book clearly demonstrates how to plan and implement TQM for any business: manufacturing or service.

We can illustrate the synergy between project management and quality with an analogy. Let's say, for example, you wanted to rewire a house. The basic requirements would be the wire, the conduit, and the electricity. The wire is needed to carry the electricity and the conduit is needed to carry the wire. It is obvious that wire is critical, but the conduit is equally important. Without it, the wire becomes difficult to route and more susceptible to damage. The result could be a short circuit or, worse yet, a fire. This is certainly not a desirable situation. Implementing TQM is very similar to rewiring the house. The house represents the company needing improvement, the wire represents the basic groundwork to succeed (communications, quality tools, training, etc.), and the conduit is project management. If the groundwork is not carefully selected, failures could occur just as if the wrong gauge wire was used in the house. The use of project management greatly reduces the probability of failure, just as the conduit reduces hazards and gets the wire to the right places. When completed properly, the house has improved electrical capacity and the company has a competitive advantage due to a pervasive quality system.

Acknowledgments

This book has been under development for about 18 months, representing a tremendous number of hours of work. There are two people to whom I would like to express my appreciation for their help.

My coauthor, Philip C. Nunn, PMP, was not initially involved in the first stages of the book. However, he enthusiastically accepted the challenge of bolstering the project management portions. He had to juggle several other professional activities along with this one, and his contributions have helped make the book truly unique and universally applicable.

Karen McQuilkin, my fiancée, best friend, and love has been a constant source of motivation for me in completing the book. She has reminded me when I have needed reminding, assisted me when I have needed assistance, and pushed me (ever so gently) when I have needed pushing. With the whirlwind of activity pulling me in so many directions, I thank the Lord for Karen.

Jeffrey S. Leavitt

I want to acknowledge the generous assistance I received during my search for a better way to communicate the concepts and practice of project management, and also the assistance I received during the writing of this book. First, I want to acknowledge the influence many of my colleagues in the Project Management Institute have had on the development of my ideas. Notably, I am intellectually indebted to Max Wideman, Alan Stretton, John Adams, David Cleland, Jim Rouhan, Dick Ryder, and Bill Moylan. My dear wife, Hilde, especially deserves recognition for her tolerance of "an author in the house." I want to thank my employer, Plan Tech, and especially Jim Bongiorno, for patience and support during the writing of this book and for assistance in the preparation of the illustrations. And I am indebted to my friend and colleague, Jeff Leavitt, for this opportunity to apply my work to Total Quality Management in this book.

Philip C. Nunn

Introduction

This book is intended for use by any professional who is charged with the responsibility of augmenting the current business practices by integrating the concepts of Total Quality. Although the book can easily be read by those not in management positions, the benefits can best be realized when applied by mid- to senior-level managers. There is a widely accepted assumption in the implementation of Total Quality which is too often ignored, however: Top management is committed to Total Quality, which means the person designated as the "Total Quality Project Manager" has not only the responsibility, but also the authority. Lower levels of management or even non-management personnel may have the energy and knowledge to implement Total Quality, but top management will not relinquish the required authority.

We have organized the book to be read starting from Chapter 1 to Chapter 10. You are probably asking yourself, "Isn't that how books are usually read?" Well, yes, of course. However, many times readers will select a portion of a book which is most important to them, skipping other areas. *Total Quality Through Project Management* cannot be used in that fashion, at least not initially. The chapters are designed to be read *in order.* It is like trying to read a novel and skipping some chapters. Confusion is certain when, in Chapter 14 of the novel, Bob inherits $2 million even though it was stated in Chapter 10 that Bob was not eligible to inherit any money. Something happened in Chapters 11 through 13, and now you have to go back and read them anyway. The same concept holds true here.

Chapters 1 and 2 give the reader a perspective on Project Management and Quality Management, respectively. Both chapters cover a history of the discipline along with our viewpoints on each.

Chapters 3 through 10 are individually organized in two parts. The first part of each chapter describes the sequential step in the project management process. This includes not only the general concept for the step, but also the specific project management tools/techniques applicable within that step. The second part of each chapter applies the project management step to the implementation of Total Quality. A translation from project management to quality management terms is provided, along with a specific example of the application.

Total Quality Through Project Management

Chapter

1

Concepts of Project Management

Project Management History

Down through recorded human history there have been projects and project managers. However, that does not assure that there was project management. The pyramids were built by the Egyptians. The Romans built a road system in Europe. The English built a navy. All of these were obviously projects and, equally obviously, had project managers. What separates them from modern project management is the lack of a common systematic method of management. All three of these examples occurred at different historical times. All three were performed in different places. All three were accomplished by people with different languages and different cultures. They are unlikely to have benefited from each other's experience.

In contrast, today project management is a well-defined general process which is applied similarly throughout the world. Modern project management started as a defined process in the 1950s when the U.S. government used it to control the development of complex weapons. The dissemination of this process was haphazard. In 1968, several people outside of the government who wanted to apply the project management process met and formed the Project Management Institute. Its first purpose was to act as a forum for the discussion and exchange of ideas and experiences. This desire was facilitated through a membership list and annual symposia. By the mid-1970s it was apparent that this was a different form of management with some unique techniques.

In 1981 the Project Management Institute formally recognized the development of uniform standards for management of projects as its responsibility. The first draft of a set of professional standards was published in 1983. In 1984, the first group of project managers was tested for professional knowledge, experience, and practice according to the draft standards. The successful candidates were awarded status as Project Management Professionals.

Work continued on a common definition of project management. In 1987, it was published as the Project Management Body of Knowledge. This event rounded out the requirements for establishing project management as a profession. Today there are about 10,000 members in the Project Management Institute and almost 1500 certified Project Management Professionals. Nearly 30 colleges and universities offer programs in project management at either the bachelor's or the master's level.

Still, work continues to further refine the definition of the Body of Knowledge, certify qualifying professionals, and oversee the establishment of project management curricula in colleges and universities. The application of project management to specific industries and disciplines is receiving great emphasis. This book represents one of the first documentations of the adaptation of project management to quality assurance.

Project Management and Total Quality

The concept of TQM has been around for several decades, but its adoption has been hindered by the lack of a clear way to implement it in manufacturing industries. Most managers have been looking for a simple, direct method of bolting it onto what they have been doing. By now it is apparent that the TQM concept is incompatible with business as usual.

With the realization that another way must be found to implement TQM came the question of how. TQM must be implemented with a management method which has the opportunity for continuous improvement and customer sensitivity built into it. Project management is a prime candidate for being this vehicle.

Project management is a problem-solving method, and implementing TQM is a problem which fits the profile of project management. Every component of it is designed to facilitate the solving of complex problems. It uses teams of specialists. It makes use of a powerful scheduling method. It tightly tracks costs. It provides a vehicle for management of total quality. It plans for success and integrates action.

In government, project management is used mostly to manage the cost of development projects. The techniques of earned value are finely tuned, and even used to kill unproductive projects.

In manufacturing industries, project management is often mistakenly seen only as a sophisticated scheduling method. To a great extent, this myopic view can be attributed to the onslaught of software salespeople. They are everywhere all the time. However, their products are just one of the tools which contribute to success. The decision-making process of project management requires more than just a computer tool.

Project management is adopted best where there are recognizable problems of coordination. Engineers and engineering managers usually recognize these. They can visualize the benefit of modeling the flow of work. They are accustomed to working in teams with other specialists to develop products and processes, and to solve problems. They understand the value of having a realistic plan to follow.

These patterns of behavior were not imposed. They grew up out of necessity. Engineering is a problem-solving profession. They are interested in any method which will enhance the effectiveness and efficiency of their work. Project management was invented by engineers.

Project management has low acceptance among those professions which are not problem-oriented because they do not have the types of problems that project management can attack. Accountants, schedulers, timers, and production managers are not attracted to project management because their objectives are different or are only a part of what project management requires. They are concerned with the speed of repetitive work, or they are groomers of data for reports to higher managers.

Even in mass production operations where the management tools are finely tuned to the operations, project management is used to make changes efficiently. Model changeovers, new process introductions, construction and tryout of new tools, and new facilities are all managed with project management.

Project management makes changes. By definition, projects have a start, work accomplished, and a finish. The finish comes when the objectives for the project are satisfied. Project objectives always address changes which will be made in some current situation. If an organization does not want to make a change, then project management is not an appropriate management method for them. This does not imply that changes should not be made there, only that there is no motivation for change. In such an organization, the introduction of project management would have little support and maybe even resistance.

The Framework of Project Management

Project management is a process which was designed to manage projects which in themselves are processes. Both Max Wideman* and Alan Stretton,† writing in the *Project Management Body of Knowledge,* see project management as having a life cycle and three major dimensions. Wideman sees the major dimensions as (1) the functions of project management, (2) the project's phases, and (3) management process. Stretton sees the three major dimensions as (1) the functions of project management, (2) the project's phases, and (3) tools and techniques of project management. Philip Nunn,‡ writing in the same book, sees project management as an integrative process. Since he is also an author here, that's the approach we'll follow. It has a more practical orientation and leads to a set of steps to follow for implementing TQM.

The functions of project management which define the scope of the profession are:

---

* Wideman, Max, *Project Management Body of Knowledge,* Project Management Institute, Drexel Hill, Pa., pp. 1-1–1-6, 1987.

† Stretton, Alan, *Project Management Journal,* Project Management Institute, Drexel Hill, Pa., August 1989.

‡ Nunn, Philip, *Project Management Body of Knowledge,* Project Management Institute, Drexel Hill, Pa., pp. 3-1–3-5, 1987.

- Scope management
- Quality management
- Time management
- Cost management
- Risk management
- Human resources management
- Procurement management
- Communications and information management

These functions are described in detail in the Project Management Institute's *Body of Knowledge.* The process applied here integrates these functions for application to a project.

Project management covers a broad field. There are bachelor's, master's, and Ph.D. degree programs in project management at more than 30 colleges and universities.

Project management is a process, not a thing or a structure. It's dynamic, changing its focus as it moves through its life cycle on a project. In practice, this means addressing problems today that are different than they were yesterday.

Project management is nonstandard. It applies to projects, and projects are nonstandard work. If they were standard work, there would be procedures; there would be fixed routines. Some organizations have tried to do that to projects, but the results have been unsatisfactory—failures. This is the management of change. Project management is the only management method with change built into it. It was designed that way.

Project management doesn't replace your current organization structure. It is not an alternative management method. It is an additional capability for managing changes whether they are development of new products, implementation of new processes, or modification of facilities. There is no need for apprehension. You are not going to be reorganized. You are adding to your capabilities.

During the execution of a project, people may work together in a different way than usual. It is teamwork. It requires the dual elements of effectiveness: empowerment of employees and involvement of managers as a team. However, it works better at some times than at others. We cannot predict the outcome of teamwork or its strength. It's situational and subject to all the vagaries of human personality.

A key factor in project teamwork is leadership, because the team is a temporary structure. Leadership and motivation are bound together for effectiveness.

Another key factor for success is the ability to negotiate for resources. In an industrial environment, these are the day-to-day negotiations we go through with other managers when we need the skill and time of their people. When the negotiation is successful, our problems are just starting. We have to motivate the people to work for us. Motivation is the method used because, if we don't own the people, we can't direct them.

Is this beginning to seem like a different management approach? Are you thinking it's going to make your job more difficult? In some respects, you're right. Project management is not easier, but it is better. Remember, the alternative is chaos, and chaos always costs more in effort, time, and dollars.

Key Concepts

Before launching into the process for applying project management to the implementation of TQM, let's review a few key concepts of project management.

Project management is the integration of several skills and disciplines. It requires a breadth of knowledge and workable understanding of the relationships between skills. These relationships must be focused on the project's objectives.

When managing a project and tracking progress, the information must flow in a closed loop. There must be feedback to the people doing the work. The primary customers of project management information are the people doing the work. Reporting to management is incidental to the team.

Project management uses software as a calculating tool. I have had a lot of people ask me for software training. In discussing what they want specifically, I almost always discover that they know how to run the software. They can run it, but they don't know how to use it. They don't know what it's telling them or why it's telling them what it is. They don't know how to use the information the software produces. They don't need software training—they need project management training. Project management software, like any other complex tool, is used within a context.

Project management requires leaders. For the head of a project, leadership skills are more important than technical knowledge. The technical experts are the team members. The leader must know enough about all of their specialities to be able to keep them talking together, but not get in their way. I have found in industrial programs that our best technical people often are not the best leaders. They are very good technicians, absolutely excellent; they're dedicated to quality work, but they don't get along with people very well. Although some brilliant people are difficult to get along with, they are extremely valuable to the project. In project management, there is a great deal of frustration. The leader must have a high tolerance for frustration.

All projects start with objectives. The objectives clearly state what the project is supposed to accomplish. These must be stated at the beginning of the project. Their accomplishment must be measurable because the finish of the project is when all of its objectives have been accomplished. Therefore, when a project is born, its death warrant is written.

A key rule to project management is that intelligent planning must be done up front. There are too many contingencies and other unknowns in a project to just start work with only a vague notion of what will be done. It's the coordination of activity that usually gets neglected. Each group may know very well what they must do, but the handoffs between them are what the manager of the program must coordinate. Up-front planning identifies these and when they are expected to occur. When the manager manages these handoffs, he or she integrates the activity of the program.

Not all work is appropriately managed by project management. It is designed for transient work (that is, work which, when completed, goes away). It has a start and a finish, and the work is guided by specific performance objectives. It usually crosses organizational boundaries so it requires special attention.

Chapter

2

Concepts of Total Quality

Quality Management History

Similar to project management, there has always been the need for quality in many areas over time. Prior to assembly-line processes, quality was the responsibility of the individual performing the work. There was usually only one person doing the work. "Pride of workmanship" was synonymous with quality, for there was only one person making the Swiss watch or wagon wheel.

The development of the assembly-line process created a whole new set of circumstances under which products were manufactured. Efficiencies became an overriding concern because quantity meant money and, likewise, poorly managed processes lost money. A new way of creating efficient systems was necessary in the evolution of the industrial age.

In the 1920s Dr. Walter Shewhart of Bell Laboratories developed a statistical technique called *control charting*. It was used to track the process and make improvements by plotting data in time sequence. Wartime in the 1940s brought about the necessity for acceptance sampling, and G. D. Edwards, H. F. Dodge, G. R. Gause, H. G. Romig, and M. N. Torrey fulfilled that need. Postwar Japan was in ruins, and two Americans (W. Edwards Deming and Joseph J. Juran) helped the Japanese rebuild their industrial base. The time from 1950 to 1970 was a period when Japanese products were known for their poor quality. However, persistence and consistency in application proved to be key factors in the Japanese turnaround, which was first recognized in the early 1970s.

In an attempt to gain lost market share, many U.S. companies began to imitate the successful Japanese by attempting to implement concepts such as quality circles. The problem was that individual components of what would eventually be called Total Quality Management could not stand alone. An integrated system was necessary but not easily understood, leaving the U.S. companies perplexed. In some instances, blame was mistakenly placed on cultural differences. Some innovative companies made progress, but most faltered and searched for new, quicker ways to improve quality. The creation of the Malcolm Baldrige National Quality Award in 1988 was intended to help companies by providing direction in terms of proven quality system requirements.

There still seemed to be a lack of consistent implementation methodology to ingrain the quality system throughout the fabric of a company. This book is the result of the evolution of quality systems and project management. Once on separate but parallel paths, quality management and project management principles have converged for use in the 1990s and beyond the year 2000.

Total Quality

There are tens of hundreds of books available dealing with the subject of Total Quality Management. That is why we are not going to spend a lot of time on the subject. What we are going to address in this book is the implementation of TQM through the use of project management principles and techniques. Therefore, a refresher of TQM and our interpretation should suffice.

I am really not sure where the term "Total Quality Management" originated, although the book *Total Quality Control* by Armand V. Feigenbaum has been around for quite some time. The "big three" philosophies of Joseph Juran, W. Edwards Deming, and Philip Crosby are very well recognized. None, however, use the term TQM. Some associated terms and acronyms now floating around the quality arena:

World-class quality	ISO standards
Zero defects	MBNQA
Empowerment	Employee involvement

SPC	Benchmarking
QFD	Customer focus

I guess if you took all these things along with their respective philosophies and put them into the quality kettle, you would have a TQM casserole after a couple of hours at high heat. But if we use Juran instead of Deming, or we do not use QFD, will our guests smile politely as they fill up their napkins? My guess is probably not.

What do we mean by Total Quality Management? It becomes clearer if the terms are defined separately. "Total" means complete, whole, entire, or all. Quality has had many definitions, but we will use a two-part definition. First, quality is conformance to specifications. This covers most quantifiable or tangible characteristics. Not every characteristic can be quantified, so we need the second part of the definition: meeting customer expectations. Together, these two parts define quality. Figure 2.1 illustrates the definition in the Quality Matrix™. Management is defined as planning, implementing, and controlling all activities or processes. We are planning, implementing, and controlling all processes in order to meet specifications and customer expectations. Great! But what about improvement, you say? Improvement is inherent in the definition because quality is dynamic. It has to be periodically gauged or measured. Figure 2.2 illustrates our quality process.

There must be evaluation criteria used to determine the strengths and weaknesses of a company in relation to TQM. The evaluation is divided into what we will call the *managerial system* and the *technical system.* The managerial system has six categories which establish the company climate or culture. Management vision/leadership is the overall driver, followed by communication, training, quality costs, motivation, and teamwork. The technical system contains 14 categories which focus on procedures and processes required for TQM, as follows:

Contract review	Inspect/test status
Design control	Nonconforming product
Document control	Corrective action
Purchasing data	Handling, storage, packaging, delivery
Process control	Quality records

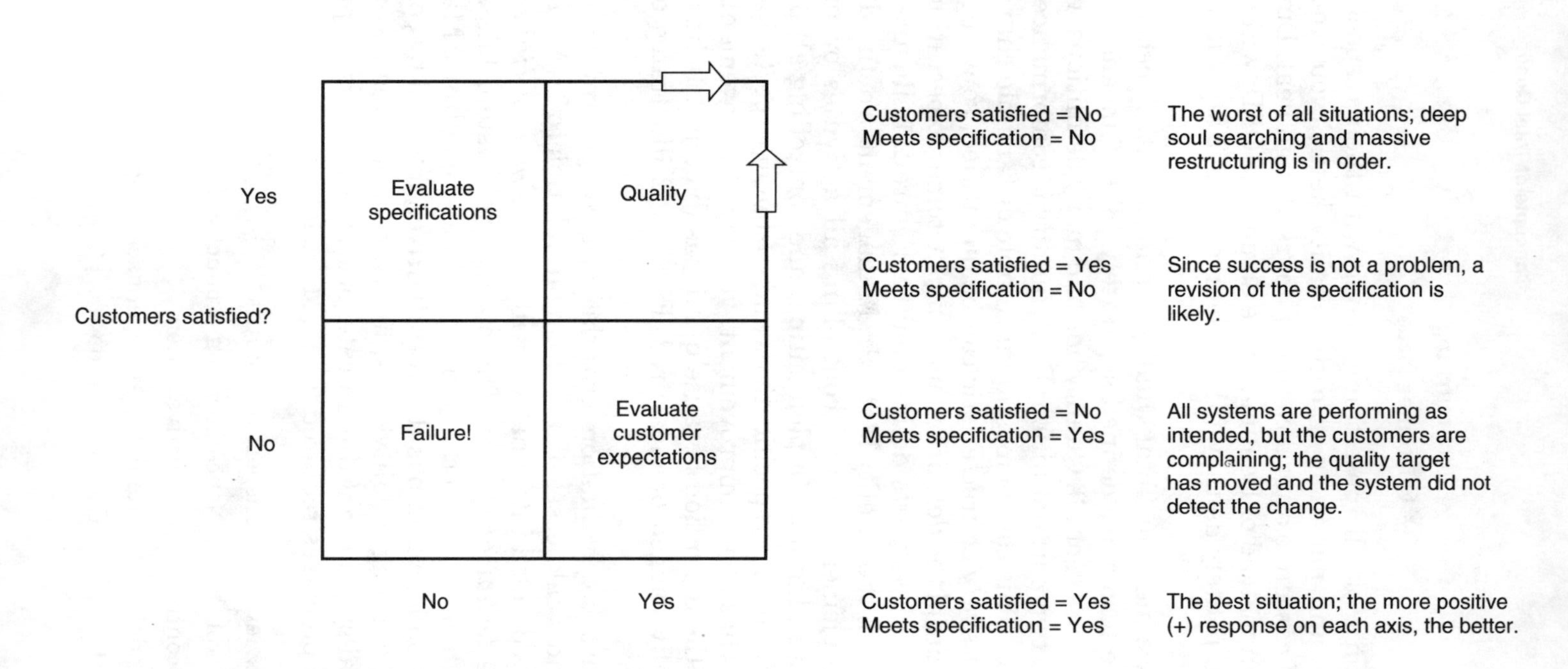
Yes
Customers satisfied?
No
Evaluate specifications
Quality
Failure!
Evaluate customer expectations
No
Yes
Meets specifications?
Customers satisfied = No
Meets specification = No
The worst of all situations; deep soul searching and massive restructuring is in order.
Customers satisfied = Yes
Meets specification = No
Since success is not a problem, a revision of the specification is likely.
Customers satisfied = No
Meets specification = Yes
All systems are performing as intended, but the customers are complaining; the quality target has moved and the system did not detect the change.
Customers satisfied = Yes
Meets specification = Yes
The best situation; the more positive (+) response on each axis, the better.

Figure 2.1 The Quality Matrix™.

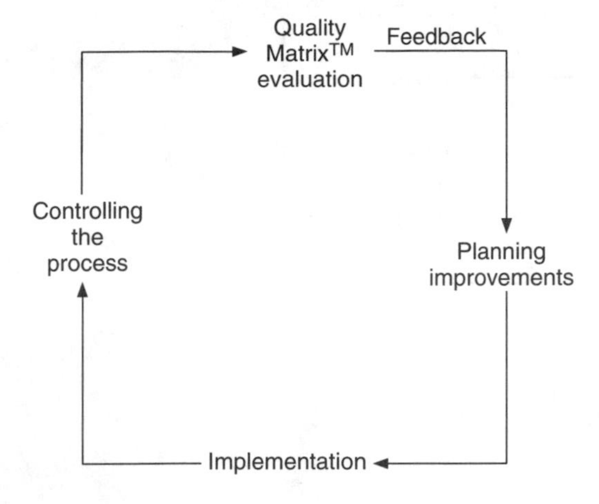

Figure 2.2 The quality process.

Inspection/test	Internal audits
Test equipment	Statistical techniques

The TQA or Total Quality Assessment is a detailed TQM survey covering the management and technical systems. Chapter 3 presents a full description along with the quantitative evaluation.

TQM can be applied to industry (manufacturing and service), government, and academia. The emphasis is on the *process,* and that might be an assembly-line process or a curriculum-review process. The complex system of processes which exist in every organization is discussed next.

The Business System

To understand how TQM operates as a company philosophy, we need to design a model from which to work. The model consists of two segments: (1) a single business and (2) related businesses. The core of the model is the single business, upon which is added the related businesses, to create the business system. The constituent parts are integrated into the business system, and can only be analyzed individually in a theoretical sense for our discussion.

The basic building block, or first stage, of our model is a single business (see Fig. 2.3). This diagram illustrates a number of key factors related to a business. The smallest unit within a business is the work center, designated WC. The work center can be

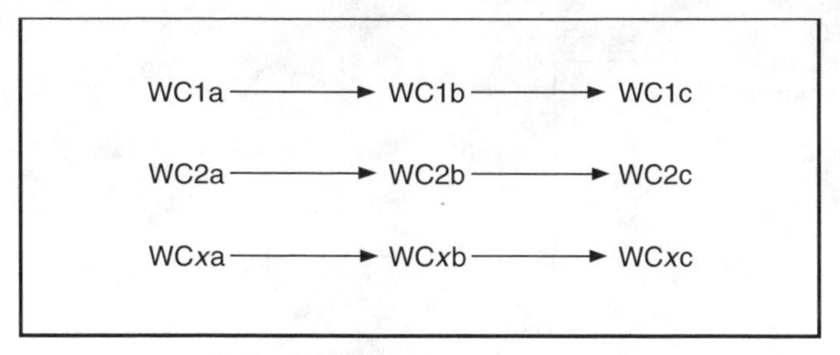

WC = Work center
1,2,...*x* = Process no.
a = Secondary supplier
b = Processor or primary supplier
c = Internal customer

Figure 2.3 A single business.

a department, a functional group within a department, or an individual. The internal customer is defined as any group or individual who receives the output (work) created by another group or individual. For the processes listed, 1, 2, and *x,* the internal customers are located at the right of the process, designated by the letter *c.* The processors or primary suppliers in this example, designated by the letter *b,* are located directly to the left of the internal customer. The secondary suppliers are at the origin of the processes with the letter *a.* The terms *primary* and *secondary* are not intended to signify relative importance. The primary supplier is the work center providing the goods or services directly to the internal customer for the specified process. The work center to the left of the primary supplier is secondary, meaning it does not provide goods or services directly to the internal customer.

Another characteristic of the single business is that the secondary supplier at the far left of the process and the internal customer may also act as interface points. If the supplier (*a*) receives input from, or the customer (*c*) delivers output to, an external source, an interface point exists. These interface points within a business have a dual functionality. They remain an internal supplier or customer while interacting with the external source. Examples of this might be a customer calling the customer service department to place an order, or the shipping department delivering a product to a customer. Customer service and shipping are at interface points with both internal work units and external sources. This is a particularly interesting example in that a *customer* can actually be a *supplier* (of information, in the first case).

The internal processes of a business become more complicated as multiple suppliers and customers are added. J. M. Juran's TRIPROL™ diagram illustrates how the work center acts as supplier, processor, and customer, as well as showing the multiplicity of input and output sources.* This condition exists in many instances, such as a manufacturing line. Many individual components from different work units are assembled into one, which is then sent to several customers. Additionally, there are usually more than three work units defining a process. Depending on the level of detail, a supplier may be secondary, tertiary, quaternary, etc. Combine these with multiple input/output sources at the origin and internal customer, and you have an intricate process.

Let's see how a simple process might look with only three work units, then magnify the process to a greater degree. Figure 2.4 is a payroll process where the employee fills out a time card, the supervisor approves the card, and accounting calculates the pay.

Normally, a supervisor will have more than one employee, so the next level of detail would include several more secondary suppliers. The supervisor might also have to provide the work information to personnel and accounting. Now there are multiple customers as well (see Fig. 2.5). A second-level approval by a manager may be required, so the process becomes elongated horizontally (see Fig. 2.6). Notice that, as the process is magnified, the roles of some work units change. The supervisor became a secondary supplier when the manager was added to the process.

We could make the process even more complicated by adding external customers beyond accounting and personnel, such as a bank and the temporary service, which would also create interface points.

The level of detail is selected by the person or group evaluating the process. It may be perfectly acceptable to define the process with only three work units. However, when implementing

---

* Juran, J.M., *Juran on Planning for Quality,* The Free Press, New York, 1988.

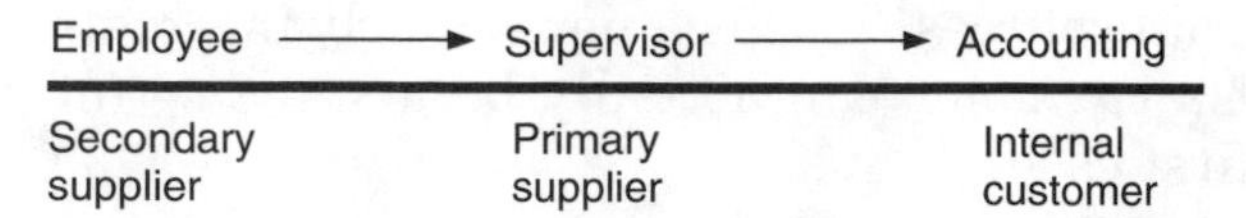

Figure 2.4 A payroll process.

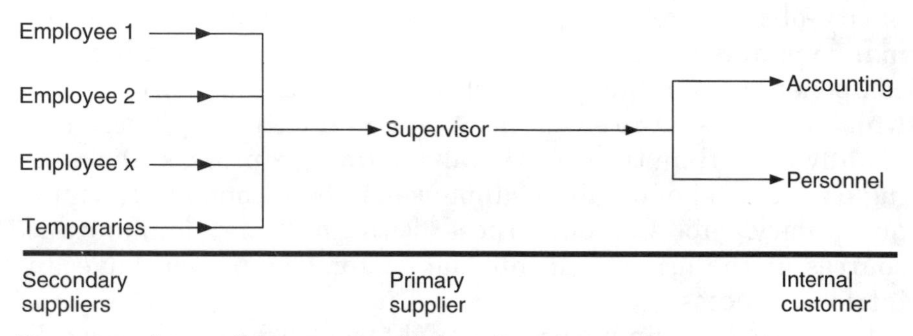

Figure 2.5 A payroll process—multiple customers.

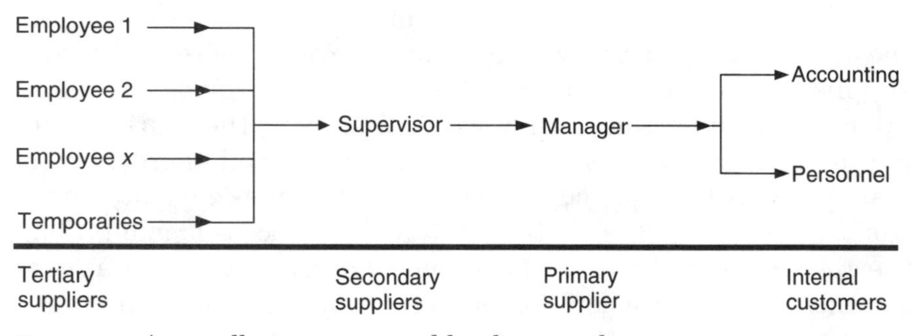

Figure 2.6 A payroll process—second-level approval.

TQM or managing a quality improvement project, a much greater level of detail is necessary.

A final comment about the single business: the number of internal processes which exist is enormous. Because of the fact that a work center will act as supplier, processor, and customer (Juran's TRIPROL™ concept), there are almost an infinite number of processes which can be examined.

The second stage of the business system is the concept of three related businesses. It is identical in form to a simple process within a single business. The difference is we are examining three businesses instead of three work units (Fig. 2.7). Virtually all of the principles discussed in relation to a single business apply to three related businesses. The businesses A, B, and C have been substituted for the work units. We could also expand the diagram in a similar manner to include several secondary suppliers and customers (see Fig. 2.8).

When the two concepts of a single business and related businesses are combined, the result is the *business system* (See Fig. 2.9). Tertiary suppliers have been added to the diagram to indicate the possible presence of suppliers upstream of the secondary suppliers. This is a common situation and is especially important in quality improvement projects as well as TQM development.

The business system is quite complex and requires a great deal of analysis to digest it. This is critical in determining the conditions for a complete TQM implementation. Not only do the internal processes of the business have to be identified and improved, but the internal processes of the suppliers and customers must also be addressed in the same manner. The controls

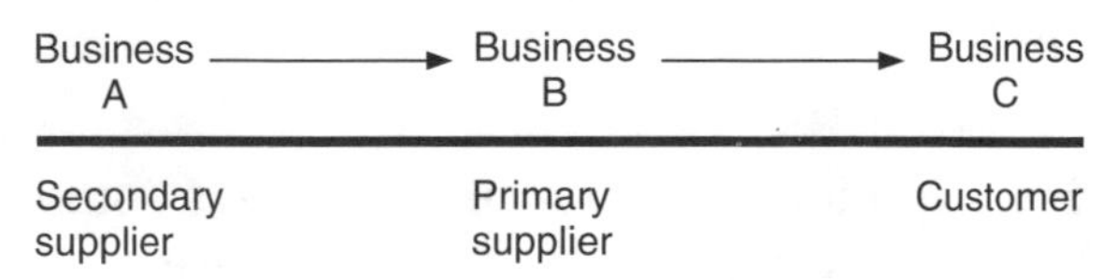

Figure 2.7 Three businesses.

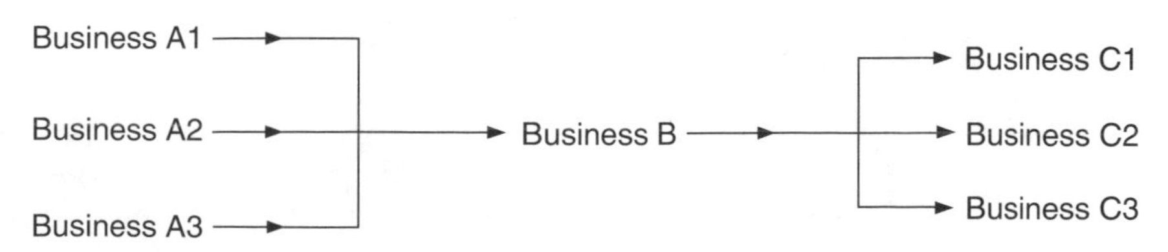

Figure 2.8 Three businesses with customers and suppliers.

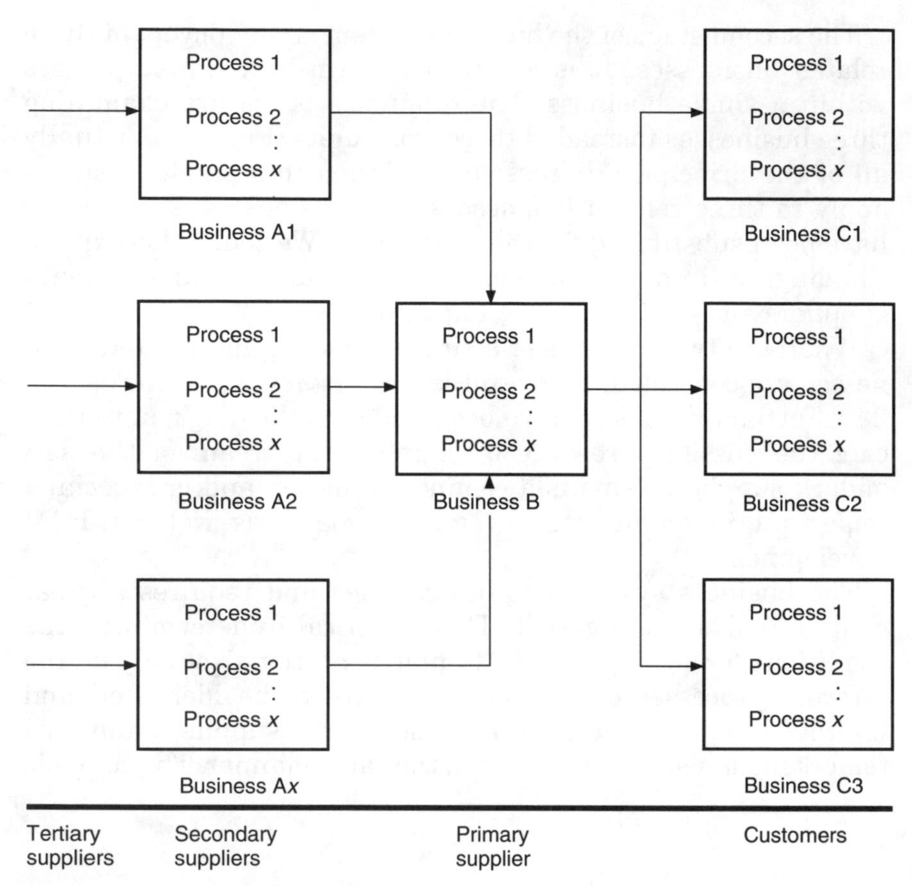

Figure 2.9 The business system.

and means for improving supplier/customer processes are somewhat different from the internal improvements, and there must be unprecedented cooperation between processors and their supplier/customer bases. You are probably beginning to see just how difficult TQM is to execute; that's why so few companies have reached "world-class quality" status.

Chapter

3

Step 1: Describe the Project

Defining the Project

We start the process by defining what we are going to do and the limitations to doing it. Figure 3.1 shows an example of the principal document of this step. This is a format rather than a document. It is frequently formatted on a computer screen so that as much information as necessary can be entered into each part. The categories of information are more important than the form itself.

Standardization of a form like this is too constraining for all but the simplest projects. Most managers need more space for information than the form would provide. The information should not be constrained to fit the space of a form.

This is a critical step, but one that beginners and novices tend to skip. Omitting the definition step can cause several identifiable problems later in the project. For example, without stated operational objectives which are agreed to, the end of the project is difficult to define. Without stated critical limitations, the scope of the project tends to grow uncontrollably. These are two of the many punishments earned by skipping the definition step.

This form is prepared before the project is started. It is usually filled in by the senior manager who will be appointing the project owner. It is filled in with the participation of the new project owner and project leaders.

This is an executive or senior management function. It defines what the sponsor wants to have done. It is the sponsor's instruc-

Project Definition Worksheet (Step 1)
(To be completed by accountable manager)

Project:

Date:

Leader:
Estimated total cost:
Requested completion date:

Project description:

Current situation:

Future state:

Strategy:

Operational objectives (includes milestones):

Critical limitations (assumptions):

Completed by:

____
(accountable manager)

Figure 3.1 Project definition worksheet.

tions to the project management team and probably supplements a memo. The incentive to take the time to formulate this definition is that the team can more accurately accomplish what is wanted.

The sponsor may ask, "Why do I have to fill this in?" The answer is, "How do you delegate work that you haven't defined?" This is only defining what is being delegated to the project owner and his or her project leaders. On this basis, there hasn't been much argument—reluctance, perhaps, but not argument. Sponsors may not want to do the work, but they don't argue that it shouldn't be done. This information can be provided in word processing format, handwritten in pencil or ink, or whatever, but it should be done.

When people give us clearer instructions, they are assuming some accountability for the results. Some sponsors may not want to risk being accountable for the results. This is the same difficulty we have with the selection of a project sponsor. Who wants to be accountable for the problems?

Project description

This is a simple description of the project, including what it is, what it is intended to do, how long it will take, and who is the accountable owner. The description of a project doesn't always come easily. Those projects which are difficult to describe are the ones which need this entire definition step the most.

Current situation

This describes the starting point for the change we are going to make. It should answer several related questions. What do we have now? What is being used now? Who is using it? When? For what? It is important that we understand that we must have a starting point. All projects build from some current starting point. Even in the story of creation, there was a current situation—chaos.

For the implementation of a TQM program, this information is provided by the Total Quality Assessment (TQA) described in the second part of this chapter.

Future state

This describes where we would like to go with this project. We can never be certain about what is going to happen in the future, but we can at least declare what we would like to accomplish with this project. What changes will we have made?

Strategy

This is the general method which will be used to accomplish the project. How will unknowns be handled? If new technology is needed, how will it be obtained? What is the organizational strategy for the project? Is it going to be pure matrix with team members staying in their home organizations? Will there be an adjacently located core team? Will there be a dedicated team or shared resources? Will contractors be used? How will the utilization of time be controlled? How will spending be controlled? How will the quality of work be determined?

Operational objectives

Operational objectives are the major objectives within the project. They are the deliverables which must be completed, and for which major milestones are scheduled. In project management, a milestone is a deliverable with a date. The sequence of developing schedule milestones is shown in Fig. 3.2. Each operational objective should address a deliverable. This is not necessarily a deliverable to a client. It may be a deliverable to someone else in the project. Each major deliverable should have a phase of the Work Breakdown Structure devoted to developing it. When the work is completed, the deliverable is received, and the quality accepted, the schedule milestone is finished. For a deliverable to be received, its quality must be acceptable to the recipient. Therefore, the milestone is not satisfied until the recipient judges the quality as acceptable. By following this sequence, our objectives get dates assigned to them.

Critical limitations (assumptions)

The critical limitations of a project are those assumptions which if not true will seriously alter the scope of the project or

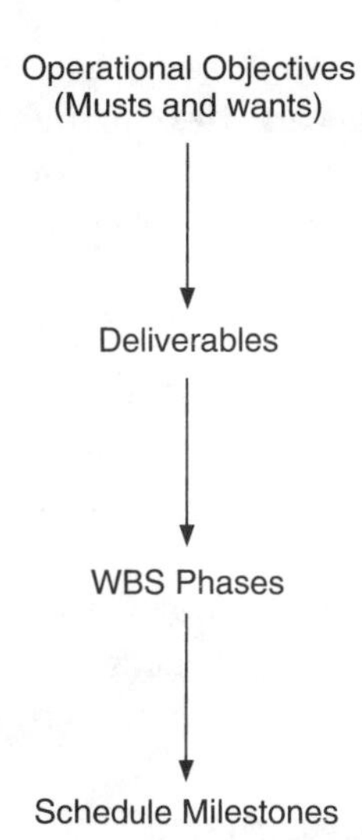

Figure 3.2 Milestone development.

kill it. Critical limitations are important. It is smart to write them down, so we have a record of what we assumed at the start of the project. This makes life simpler later when things start changing.

Discussion

Definition may seem quite elementary and unnecessary. In a well-defined project like replacing an old machine tool, it may be. But, for projects with vague definitions, it is a vital step. In fact, it is a step without which the project does not continue without a lot of misunderstanding and misdirection. I remember one project in which this step was the key that unlocked the problem.*

The management team of an industrial engineering department was trying to draft a human resources development plan. They were having difficulty deciding what it should contain.

---

* Nunn, P., "The Application of Project Management to Human Resources Planning," *1990 Proceedings: Project Management Institutes 1990 Annual Seminar/Symposium,* Calgary, Canada, pp. 201–210.

They were engineers, not human resource professionals; yet, as managers they had to develop the plan. They asked their project management consultant who was on site if he could help. They had recognized the ability of project management to clarify problems. The consultant said that he didn't know if anything like this had ever been tried, but he was willing to work with them on an experimental basis. He was depending on the analytical tools used in project management to define projects to clarify this problem. He also, incidentally, had a strong social science background.

Because this was the definition step for development of a policy, some deviation from the format suggested here occurred. TQM, however, is also a policy, so the thought process for this policy applies to the implementation of TQM.

Discussion of the project began with a general discussion of what the human resource development plan should do for the engineers, the company, and the customers. This was a broad-ranging discussion which was intended to stimulate thinking about the problem. It was constrained brainstorming.

One of the topics which had to be decided was whether this would be an integrated policy or a set of related policies, such as training, career development, and promotion. It was decided that more would be gained if an integrated policy could be developed. If this proved to be too difficult, then separate policies would be developed as a second choice.

The concept of an integrated human resource development plan focused on the longitudinal career of an engineer. The overall goal was to retain an engineer for the rest of his or her working career. For recent graduates, this could cover a span of 45 years. Therefore, the plan would address needs in three phases of an engineer:

1. Recruiting
2. Retaining trained engineers during their productive years
3. Motivating older engineers nearing the end of their careers

This concept is shown in diagrammatic form in Fig. 3.3. The activities related to these career phases, which needed to be addressed in the plan were:

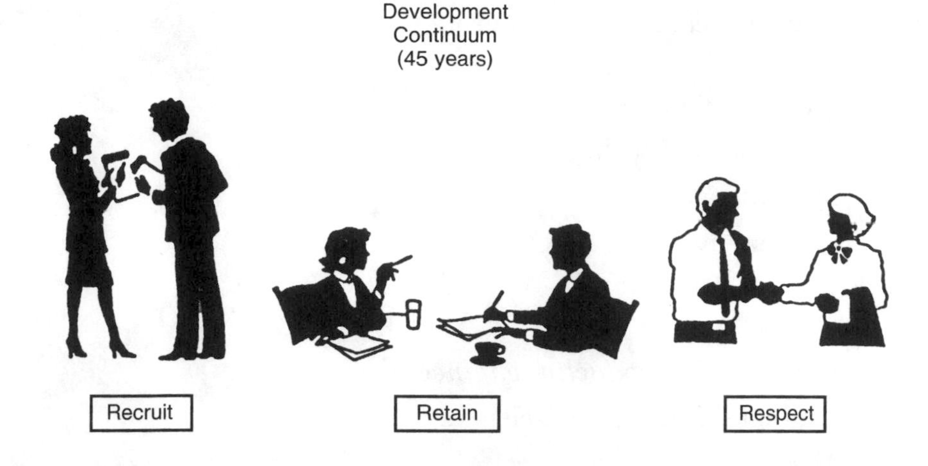

Figure 3.3 Professional development continuum.

1. Recruit
2. Retain
3. Respect

The diagram shows that these activities are time-phased over an engineer's career.

The methods of development as time progresses are:

1. Publicity to attract candidates
2. Training to focus the capabilities of new engineers
3. Mentoring to guide new engineers and keep experienced engineers active and productive
4. Promotion to provide engineers with a visible sign of achievement

Specific motivational factors were identified for each stage of an engineer's career. These were:

Motivators for recruits

- State-of-the-art technology
- Continuing education
- Expert mentors
- Fun

Motivators for productive engineers

- Meaningful impact on decisions
- Access to management
- Professional responsibility and freedom
- Money

Motivators for older engineers

- Recognition as an expert
- Respect by peers, company, and profession

Operational objectives were explored as items which must be in the plan. They identified the following objectives:

- Comply with legal requirements.
- Implement with minimum disruption.
- Facilitate anticipatory planning.
- Require goal setting.
- Require planned assignments with clients.
- Provide criteria for promotion/assignments.

Because no business is immune to economic fluctuations, the plan would have to be developed under some assumptions. These were the critical limitations. If they changed or disappeared, parts of the plan might become invalid. The management team identified these assumptions:

- There are no major budget constraints.
- The engineers want the plan.
- Clients want the plan.
- The plan is generally needed.

This concluded the initial generation of the project definition. It was developed in a three-hour session. The results were organized, documented, and distributed so they would be available for the next meeting, which would be devoted to identifying the major divisions of the plan in work breakdown structure form.

Definition is a critical step in implementation of a new policy like TQM because, to some in the organization, it will be a major change in the philosophy of their jobs.

Describing the TQM Project

The first critical step in implementing TQM is the project description. We must have a very clear understanding of where the organization is and where it will ideally be after the implementation. The Total Quality Assessment (TQA), or audit, is a prerequisite of the plan which will guide the efforts of the organization. The TQA is as necessary as a house remodeling survey. Having your house remodeled is a common activity, but what if only the kitchen is in need of repair? A more detailed analysis might reveal that only the walls of the kitchen need painting. It would certainly be a waste of resources to do anything else. Similarly, the TQA ferrets out the necessary from the unnecessary and becomes the foundation for the remaining TQM project activities.

In this section we will discuss several components of the project description. Determining the current status of a company in relation to TQM is shown. The Malcolm Baldrige National Quality Award and the ANSI/ASQC Q91-1987 standard together form the model for our TQM system. Although there is some crossover of areas, the MBNQA is predominantly managerial in nature, while the ANSI/ASQC standard is technically oriented. The future state and strategy is the subsequent task, followed by operational objectives and assumptions.

Current status

The current status of the business system is divided into three major areas: (1) elements which are in place, (2) elements which are not in place, and (3) elements which are in place but need improvement. The best approach is to perform Total Quality Assessment (TQA). The TQA is designed to compare current practices to those required for TQM.

The assessment of the business system occurs both vertically and horizontally. Expressed as a tree diagram, the business system would be at the top (level 1), followed by internal and external components (level 2). At level 3 the internal portion extends into each of the functional areas, while the external portion is comprised of the customer and supplier. Figure 3.4 shows the tree diagram down to the third level. Some companies have more functional areas than are illustrated, so that will have to be taken into consideration on an individual basis. Also, the functional areas could be divided into subfunctions. For instance, manufacturing may consist of a maintenance department and several production departments. To express the concept of TQM implementation as universally as possible, the business system detail will not exceed level 3 and six functional areas. To assure a complete assessment, it is recommended that the tree diagram be created to a sufficient level of detail.

Since the tree diagram shows only the location of the evaluation, what is to be evaluated? Previously mentioned were the managerial system and the technical system, which make up the TQM system. The managerial system consists of six sections, which cut across all internal and external lines. The audit must determine how effectively vision/leadership, communication, training, motivation, teamwork, and quality costs have been deployed throughout the business system.

A sample of all functions, both internal and external, is required for the audit. It begins with top management and follows through the various levels. The evaluation of a particular section would stop at whatever level the evidence would cease to exist. For example, if top management acknowledges the lack of a quality cost system, it would be pointless searching for evidence of it at lower levels. The following questions represent most of the items which need to be addressed during assessment.

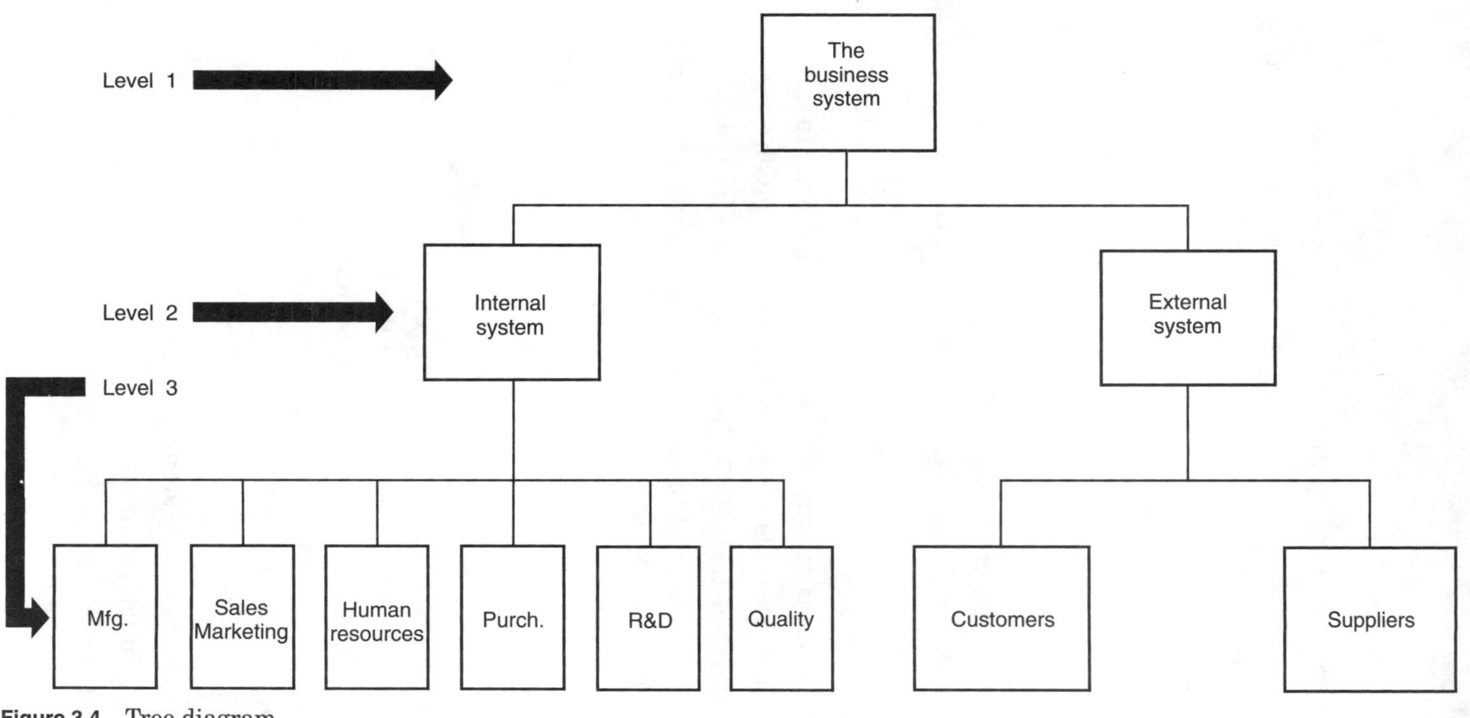

Figure 3.4 Tree diagram.

Vision / leadership

1. Is there a written “quality policy”?
2. Does it address the company’s position with regard to these key areas?
   - Customers
   - Suppliers
   - Employees
   - Community/environment
   - The business we are in
3. Are all employees aware of the quality policy?
4. What are the company’s long-term strategies in relation to quality?

Communication

1. Is relevant company business communicated to employees?
2. At what intervals is the communication given?
3. By what means is the information disseminated to employees?
4. Is there a method to determine the effectiveness of the communication?
5. What are some examples (internal and external) of items which have been communicated, and to what levels?

Training

1. Who receives training in the organization?
2. Who determines which employees will be trained, in what areas, and the timetable for training?
3. How is the effectiveness of the training measured?
4. What is the training schedule, and what training has been done in the last six months?

Quality costs

1. Is a quality cost system utilized?
2. Which functions are involved?
3. What is the review process for quality cost information and how are the results used?

Motivation

1. Which methods of motivation are used in the organization?
2. How is the level of motivation determined and tracked, and how is the information used for future improvements?

Teamwork / employee involvement

1. How are employees encouraged to participate in product and service improvements?
2. How are the effects of participation tracked and used for future improvements?
3. Who is involved in team activities?

Customer relations

1. How is customer satisfaction determined and tracked?
2. How is the information on customer satisfaction used for future improvement?
3. How does the company demonstrate its commitment to customers?
4. Is the competition used as a basis for comparison? How?

Supplier relations

1. What criteria are used in the selection of suppliers or subcontractors?
2. Who is involved in the decision of supplier selection?
3. How is supplier performance determined and tracked?
4. What methods are used to improve supplier performance?

The other half of the TQA consists of the technical aspects. Unlike the managerial portion, which cuts across the entire business system, the technical aspects are more closely related to a specific function. For instance, accounting would not be involved in the gauge calibration procedures.

Contract review

1. What is the procedure for reviewing contracts?
2. Who is involved in the contract review process?
3. Are there records of the contract reviews?

Design control

1. What is the procedure for design control?
2. Are there design plans? Do they identify the responsibility for each activity?
3. Are design inputs, outputs, and verifications planned and documented?
4. What are the procedures for design changes?

Document control

1. What is the procedure for document approval, issue, and change?

Purchasing data

1. Do purchasing documents contain data clearly describing the item being ordered, such as title, grade, and specification reference?
2. What methods of traceability are used for a purchased product?

Process control

1. How are the processes planned and controlled to assure quality?
2. Are work instructions and workmanship criteria available where the work is being performed?

Inspection and testing

1. What measures are taken to assure that incoming product meets specified requirements?
2. What are the provisions for incoming product which is released for urgent production purposes?
3. What are the procedures for in-process inspection and testing?
4. What are the procedures for final inspection and testing?

Inspection, measuring, and test equipment

1. What procedures are in place to assure proper calibration of equipment?
2. What standards are used for verification?
3. How is the capability of test equipment determined?

Inspection and test status

1. How is the status of product relating to inspection and test results identified through the process?

Nonconforming product

1. What procedures exist to identify and segregate nonconforming material?
2. Who is responsible for reviewing and authorizing the disposition of nonconforming material?

Corrective action

1. What are your procedures for corrective action?
2. How do you determine if the corrective action is effective?

Handling, storage, packaging, and delivery

1. What procedures exist to prevent damage or deterioration of product in handling and storage?
2. How is product controlled to ensure identity and segregation from other items?
3. What measures are taken to assure quality during delivery to customer?

Quality records

1. What records are kept and for what length of time?
2. Are quality records easily retrievable?

Internal quality audits

1. How often are internal quality audits performed?
2. What is the schedule for internal quality audits?
3. What mechanism exists to ensure timely and effective corrective action on nonconformances?

Statistical techniques

1. What statistical techniques are used to monitor and improve processes?
2. How are different techniques selected and applied for control and improvement purposes?

To assist in the TQA, a list of documentation requirements and record requirements are provided.

Documentation

1. Quality policy
2. Final inspection and test
3. Instrument calibration
4. Control of nonconforming product
5. Corrective action
6. Handling, storage, packaging, and delivery
7. Internal quality audits
8. Design control (includes inputs, outputs, verification, and design changes)

Records

1. Management review
2. Assessment of suppliers/subcontractors
3. Instrument calibration
4. Internal audits
5. Training
6. Contract review
7. Design verification
8. Product identification
9. Inspection and testing
10. Responsibility for product release
11. Nonconforming product

As a further aid in determining the current status using the TQA, point values can be assigned to the managerial and technical systems of TQM. The management system consists of eight categories, while the technical system has fourteen. Each category can be assigned a total of three points, one for each of the following:

1. Is the category element evident?
2. Do all appropriate functions/levels understand and make use of the category element?
3. Is the category element functioning as intended?

These three broad questions are structured so that no points can be awarded if the previous question is scored 0. This gives us a quantitative view of the current status with 24 points for the managerial system and 42 points for the technical system. We can then convert to percent acceptable by dividing actual points by the points available for each system. World-class status would have a minimum of 90 percent for each system, with no individual category being less than two points. The position of the company can be graphically illustrated on the TQMatrix™ (Fig. 3.5). Two numbers are used to mark the overall location and the quantity of categories receiving less than two points. The next task is to determine where we want to go, and how we want to get there.

Future state/strategy

Just as you need to decide your destination when you take a vacation, you need to decide what the company aspires to be with regard to quality. *Warning:* This is a top-management decision. Advice can be given by lower levels of management, but the ultimate decision rests with the executives.

A review of the TQA is held with the executive level, and the wheels begin to turn. Pass an ISO series standard audit? Win the MBNQA? Become a world-class company? Create an expanding job base? The future state of the company may include any of these or any other relevant macro goals. Remember, though, it is decided by top management as part of the overall strategy.

Another factor involved in developing the strategy is the structure of the organization. Will we change any or all of the organization to accomplish our goals? Will we need an outside consultant, or do the resources exist internally?

Operational objectives, deliverables, and milestones

Operational objectives are activities within the TQM project which fall into one of two categories. The first is the "must" cat-

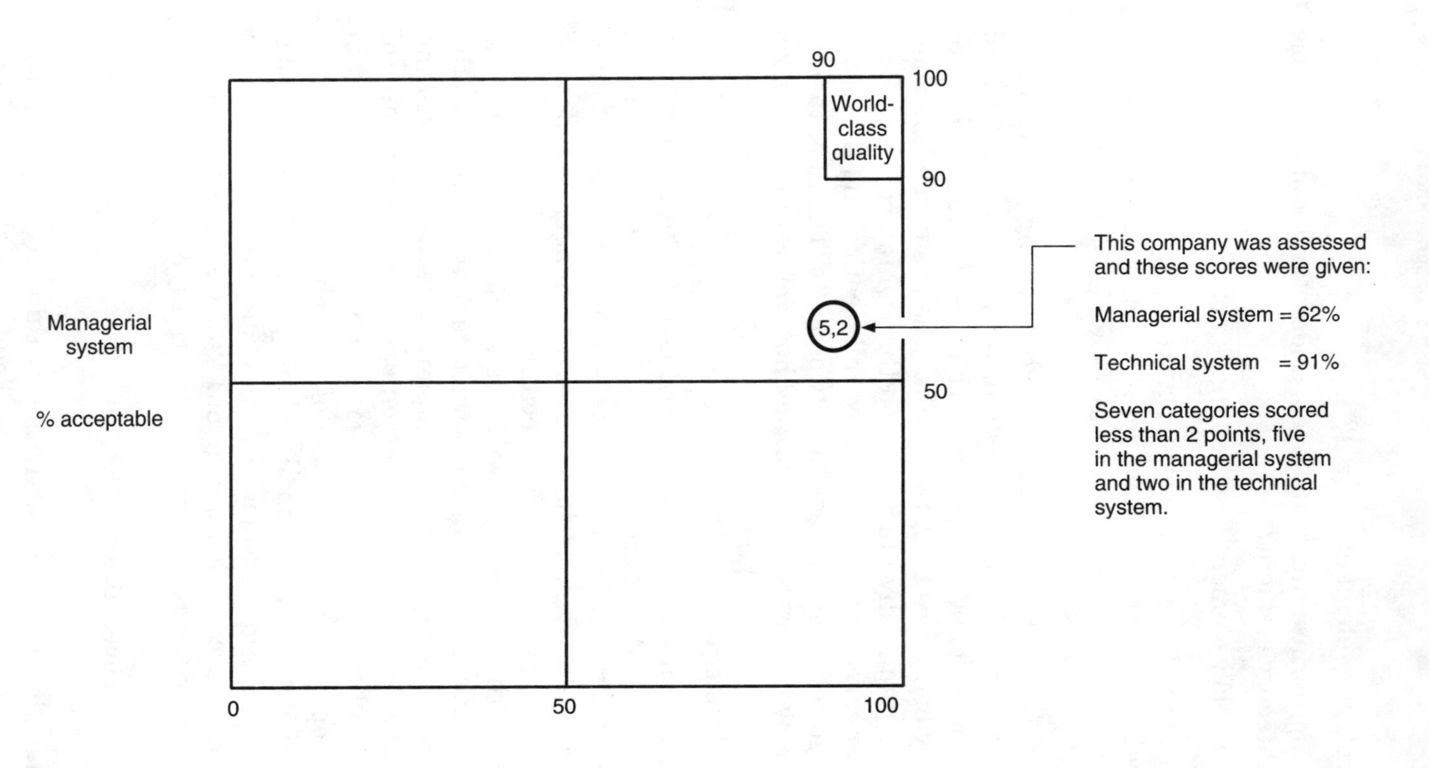

90
100
World-class quality
90
5,2
50
0
50
100
Managerial system
% acceptable
Technical system
% acceptable
This company was assessed and these scores were given:
Managerial system = 62%
Technical system = 91%
Seven categories scored less than 2 points, five in the managerial system and two in the technical system.

Figure 3.5 The TQMatrix™.

egory. In this situation, there is no alternative. "Wants," on the other hand, are optional. The success of the project will not be jeopardized if a "want" is not completed. Consider the following example in sorting wants from needs. Company XYZ does not have a formal, written quality policy.

Objective (must): Create a quality policy.

Objective (want): Write the quality policy on 30 × 45 inch blue-linen paper.

The important item here is to write the quality policy. The implementation of TQM does not hang in the balance if the policy is written on another medium.

Each operational objective requires what is known as a *deliverable*. A deliverable is the final product of the objective. It may be a document, a report, a procedure, a level of knowledge, etc. In the example of the quality policy, the deliverable is the actual document.

A schedule milestone is also associated with the operational objective. The simplest definition of a milestone is a deliverable with a date attached. Let's again refer to the objective, "create a quality policy." If the quality policy was to be written by December 31, this would be the schedule milestone.

Although very simple to do on an individual basis, the creation of operational objectives for an entire TQM project can be quite complex. If you are starting a new company, virtually none of the managerial and technical systems are in place. In most cases, however, the company will have been in existence for some time. The TQA will uncover many elements not in place and many needing improvement. The final number of operational objectives may seem almost insurmountable, but the exercise is absolutely essential.

Critical limitations (assumptions)

Critical limitations are those items, internal and external, which are assumed to be correct. But, more than that, if they are not true, the nature and possibly the success of the project is changed. The following list represents some generally accepted critical TQM assumptions, but each implementation must be evaluated on its own merits.

1. *Top management is committed to TQM.* This means that all aspects of the business system, vertically as well as horizontally, are required to participate. TQM must be presented as a "win-win" situation and long-term development.
2. *The company is being operated under sound management practices.* Recently, TQM has become the scapegoat for the demise of poorly run businesses. No TQM implementation can overcome unchecked spending, unethical behavior, or regularly scheduled leveraged buyouts. TQM is not a quick fix for these or any other problems.
3. *The company's strategic quality plan and objectives are understood by everyone.* The captain of a ship proudly exclaims, "Set Sail!" to the eager and seaworthy crew. One of the crew members, with a puzzled look, questions the order, "But, Captain, where are we going?" The captain replies quickly and with authority, "I don't know, but we must go, now."

   The *S.S. Quality* will never reach its destination if the crew operates under poorly conceived plans and nebulous direction.

The company we will use as a case study is a small-to-medium-size manufacturing firm. The TQA was performed and deficiencies were found in the following areas:

1 point	Internal communication needs improvement; no formal system in place
0 points	No use of statistical methods; system not evident in manufacturing

All other categories were determined to be acceptable.

The future state of the company, as stated by the president, is to be a world-class manufacturer in the industry. Currently positioned in second place with regard to market share, they would like to attain the number one spot. Their belief is that a superior product will be manufactured through the implementation of statistical techniques. Also, productivity will increase through more effective communication.

Operational objectives, deliverables, and milestones were developed by upper management in a brief but effective meeting (remember, this is only a case study).

Operational objective	Deliverable	Milestone
Must reduce variation in the two key manufacturing processes	Statistical data showing reduction	12/31/93
Improve communication to all employees regarding company activities	Surveys showing that employees are satisfied with communication; 90% minimum is the target	3/31/94

Two assumptions or critical limitations are involved with the success or failure of the TQM implementation. First, the increase in market through variation reduction assumes that the reason for their current position is in the manufacture of the product. If the product is not meeting customer needs, variation reduction will be an exercise in futility. Second, the assumption is that the employees are not currently satisfied. The initial survey will quickly evaluate the situation.

Chapter

4

Step 2: Appoint the Planning Team

Elements of the Team

In projects with the large scope of TQM there are usually three levels of management. They are the project sponsor, the project owner, and the project leaders. Figure 4.1 shows the organizational diagram of such a project. Project leaders probably are the most common of these positions. They are the people who manage the day-by-day work of the project. The project owner is the manager who must coordinate the activities of the subproject teams. This coordination is usually exercised through the project leaders of the teams. An important item of this coordination is the management of handoffs between the subproject teams. The position of project sponsor, although very much a part of the project, is an executive or senior manager in the permanent line organization who represents the project in executive discussions and policy decisions.

Project sponsor

The project sponsor is most often the person who started the project definition. This is also the manager who is instrumental in the selection of the project owner. These two people must be able to work together closely. They will have to agree on objectives and strategy.

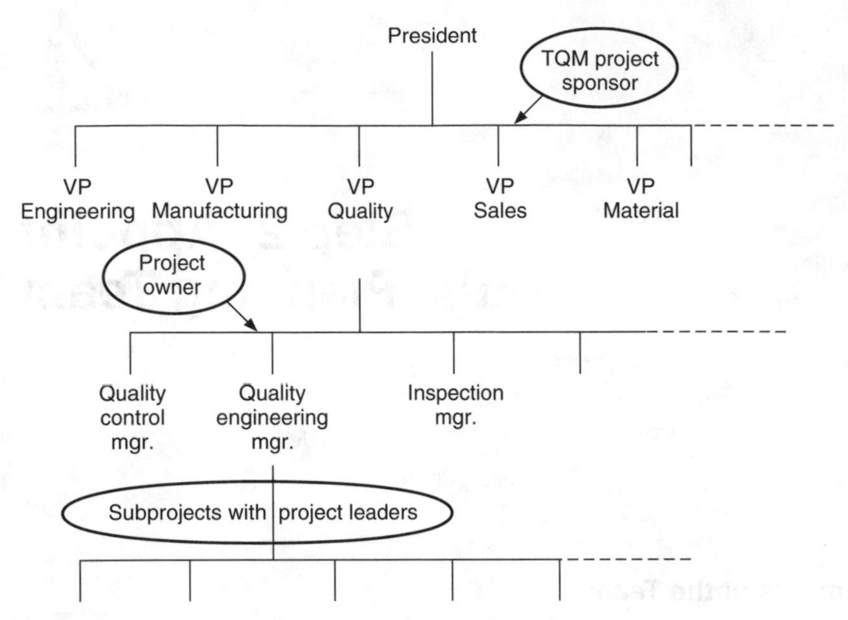

Figure 4.1 Project management organization.

The project sponsor defines the working relationship between the project leaders and the permanent line organization. This definition is necessary so the line managers as well as the project leader know what is expected of them in the matrix organization of this project.

The project sponsor, working with the project owner, defines the first level of the work breakdown structure. This is an expression of part of the strategy for the project. It defines the most summary level of reporting which will be made to management. It also defines the major blocks into which the project is divided and the organizational location of the major milestones. There is more description of the choices for work breakdown structure in Chap. 5.

The project sponsor defines the performance expectations and how they are to be measured and evaluated. Performance should be evaluated in a way consistent with company policy. This applies to the performance evaluation of the project leader as well as the project team.

The project sponsor establishes the periodic reporting requirements. This helps standardize project reporting and still allows for special requirements unique to a project.

An important, although hopefully not a regular, task of the project sponsor is to try to remove roadblocks and other impediments to the progress of the project. Indirectly, the project champion can be most effective in this task by just keeping the project visibility as high as possible. This keeps the perceived priority of the project high. Most delays in a project are the result of other work being given a higher priority in the line organization. The project sponsor is a key player in the priority game.

Project owner

The project owner is the manager who is accountable for performance of the project. He or she is usually the manager of the home organization of the project, that is, the organization which owns the project.

In a project with the scope of TQM implementation, the work is done by subproject teams. Each of these is led by a project leader. The principal responsibility of the project owner is to coordinate the activities of the subprojects. This is usually accomplished by acting as the leader of the management team comprised of the project leaders of the subprojects.

The primary focus of the project owner's coordination activity is assuring that handoffs between subprojects occur on time and that the quality of the deliverable meets expectations. The definition for the handoff deliverable is the combination privilege and responsibility of the receiving subproject. If they receive what they specify, the quality is judged as good. This is the familiar supplier-to-customer concept where the receiving subproject is the customer.

Another responsibility of the project owner is the timeliness and quality of the overall product of the project. This responsibility may require some staff which is not a part of any subproject team.

One caution for project owners is that they should try to avoid managing work within the subprojects. The project leaders for the subprojects were picked to do that job, and the team members are expected to be skilled at their work. Let them do it, but be sensitive to resource shortages which could hamper the project.

The position of project owner does not exist in small projects which have only one project team. Small projects have a project sponsor and a project leader.

Project leader

Once the project has been defined, the owner needs a project leader to run it. What type of person is this? The answer lies in what is going to be required of this person. On some projects, it seems like a superhuman individual is required. Fortunately, most projects have only a few crises; however, the leader must meet a steady flow of challenges.

The project leader must always manage tight schedules. There is never quite enough time to complete everything. All good projects have pressure on the time allowed, so the project leader must be comfortable working under schedule pressure.

Project leaders must work smoothly in an environment of multiple demands for attention. Projects contain a lot of simultaneous work, so there is constant demand for attention from several areas at any one time. These demands must be immediately arranged in some priority order for attention. Systems for quickly assigning priorities usually use a three-tier arrangement. The infamous military triage system is one that can be applied in a manufacturing company. If it is an all-out crisis, leave it alone, but offer resources. The most effective people at fighting crises are those located where the crisis is happening. If it is an important matter which can affect future work and there is time to correct it, react. If it is a small problem in a noncritical area, make sure someone is following up on it.

The project leader must always keep work focused on the project goal. In development projects there is a temptation for attention to wander toward an item which is interesting, but not directly related to the project goal. In some organizations, political considerations have a tendency to overwhelm the project goal. This condition must be confronted as soon as it becomes visible. The project leader must stay goal-oriented.

Matrix organization

Matrix organization is the name given to the concept where people who are in permanent line organizations are assigned temporarily to a project team. If such people are drawn from each functional department, the project team contains a cross section of skills from the organization. Such a project team is frequently referred to as a cross-functional team. In various companies, these are referred to as quality circles, employee involvement groups, program module teams, customer focus teams, or some other team designation relevant to the business of the company.

Matrix organization is an old idea that goes back to the early days of project management. It is usually represented by a diagram similar to the one shown in Fig. 4.2, which typically lists functional departments across the top and projects down the left side. The names of the people temporarily assigned to a project would appear in the boxes at the intersection of their department columns and the project row. Here, we'll just refer to them as project teams. Usually, the name of an individual will appear in more than one box, because most industrial companies do not have a large enough staff of professional and technical people to allocate one to each project. This situation is referred to as sharing resources.

How many projects a person can handle effectively has been studied by McCollum and Sherman.* They surveyed 64 companies which made significant investment in research and development. Forty-four of those used shared resources in their project matrices. They evaluated the effectiveness of shared resource matrices with five performance measures:

- Return on investment
- Five-year rate of growth in sales
- Patents in last three years
- Meeting schedules

* McCollum, J.K., and Sherman, J.D., "The Matrix Structure: Bane or Benefit to High Tech Organizations?," *Project Management Journal,* vol. 23, no. 2, June 1993.

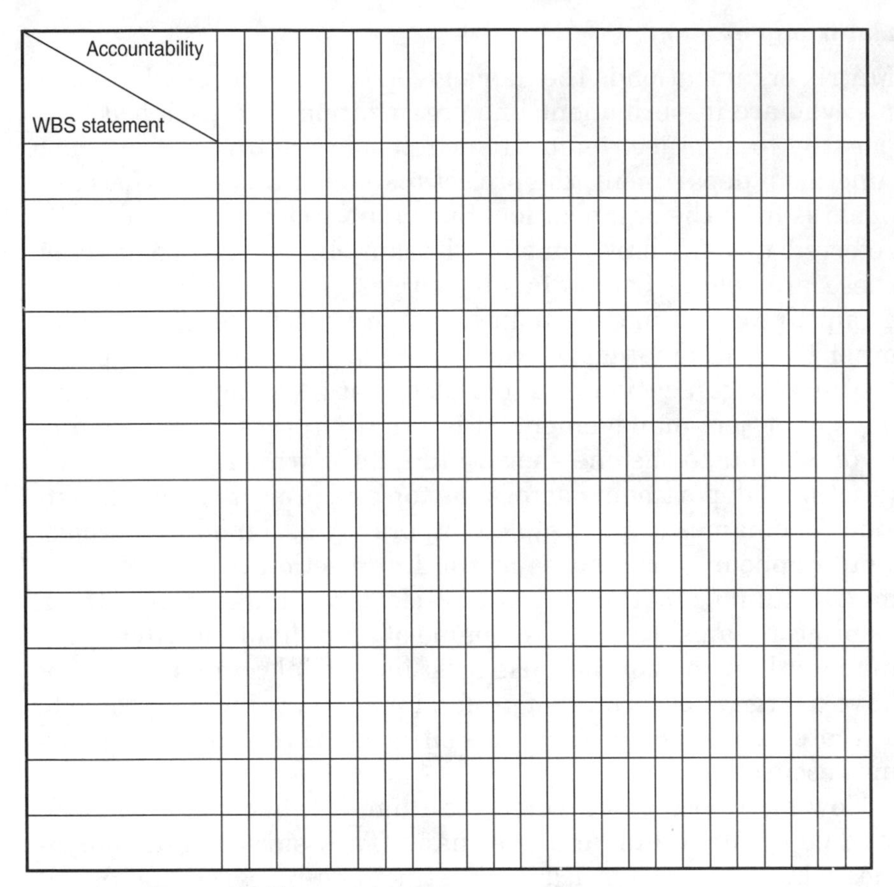

Figure 4.2 Accountability matrix.

- Projects over budget
- One-year growth in new contracts

Their results were counterintuitive. They statistically showed that the most effective project matrix of people was when people were assigned to two projects simultaneously. This evidence is contrary to the popular management belief that assigning people to only one project would be the most effective. At more than two simultaneous projects, the effectiveness drops sharply.

This study helps explain the difference between the experience of project management professionals and the unsubstantiated belief of many managers and some management writers. One cause of the disagreement is that matrix organization does not work by itself. It must be made to work. It takes effort from everyone involved with the project to make it work, and it is the responsibility of the project leader especially to make it work. If left passively to work on its own, it could lead to confusion. Misimplementation by managers where they did not encourage the teams to form could cause poor performance. Size of the team is also suspected as a cause for some poor performances. There may be a project team size beyond which matrixing people is no longer effective.

Matrix organizations theoretically can come in various strengths.* Strength of the matrix is usually referenced to the decision-making strength of the project relative to the functional departments.†

Gobeli and Larson‡ assessed the use of five different strengths of matrix as follows:

1. *Functional.* The project is divided into segments and assigned to relevant functional areas and/or groups within functional areas. The project is coordinated by functional and upper levels of management.
2. *Functional matrix.* A project manager with limited authority is designated to coordinate the project across different functional areas and/or groups. The functional managers retain responsibility and authority for their specific segments of the project.
3. *Balanced matrix.* A project manager is assigned to oversee the project and shares the responsibility and authority for completing the project with the functional managers. Project and functional managers jointly direct many work-flow segments and jointly approve many decisions.
4. *Project matrix.* A project manager is assigned to oversee the project and has primary responsibility and authority for com-

---

* Stuckenbruck, L.C., "The Matrix Organization," *Project Management Quarterly,* vol. 10, 1979, pp. 21–23.

† Galbraith, J.R., *Designing Complex Organizations,* Addison-Wesley, Reading, Mass., 1973.

‡ Gobeli, D.H., and Larson, E.W., "Relative Effectiveness of Different Project Structures," *Project Management Journal,* vol. 18, no. 2, June 1987, pp. 81–85.

pleting the project. Functional managers assign personnel as needed and provide technical expertise. This is the form of project matrix most often used by the companies surveyed by McCollum and Sherman.*

5. *Project team.* A project manager is put in charge of a project team composed of a core group of personnel from several functional areas and/or groups, assigned on a full-time basis. The functional managers have no formal involvement.

They found in their study that companies which develop new products tend to favor the functional matrix strength, while companies which develop new processes or services tend to favor the project matrix strength. The implication in this study for the implementation of TQM is that the company management will likely favor a functional matrix, while the quality management will favor a project matrix organization. Where this difference of viewpoint emerges, the final structure will have to be negotiated with the company management by the project owner and the project sponsor. This negotiation will allow everyone involved to recognize the relative strength of the project leaders.

Organizational culture

Differences in opinion about how much strength should be allowed the TQM project leader are due to differences in the subculture of different parts of the company. To cope with this type of difference, we have to have a working knowledge of cultures in manufacturing companies. In 1988, Michael Elmes and David Wilemon† published a model of organizational cultures. Their model is useful to project leaders in identifying cultures and developing a strategy for working with them. It is summarized in Fig. 4.3.

* McCollum, J.K., and Sherman, J.D., "The Matrix Structure: Bane or Benefit to High Tech Organizations?," *Project Management Journal,* vol. 23, no. 2, June 1993.

† Elmes, M., and Wilemon, D., "Organizational Culture and Project Leader Effectiveness," *Project Management Journal,* vol. 19, no. 4, 1988, pp. 54–63.

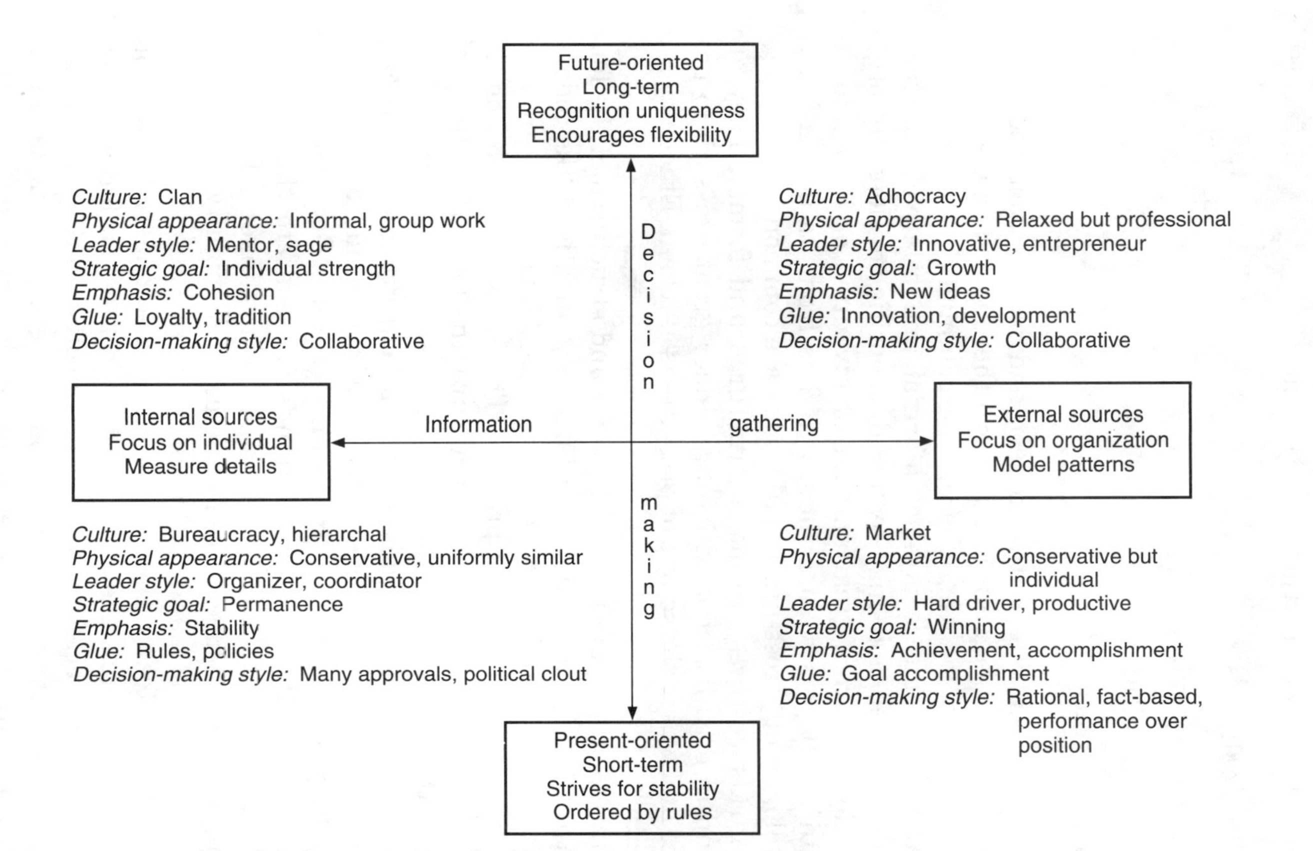

Future-oriented
Long-term
Recognition uniqueness
Encourages flexibility
Culture: Clan
Physical appearance: Informal, group work
Leader style: Mentor, sage
Strategic goal: Individual strength
Emphasis: Cohesion
Glue: Loyalty, tradition
Decision-making style: Collaborative
Culture: Adhocracy
Physical appearance: Relaxed but professional
Leader style: Innovative, entrepreneur
Strategic goal: Growth
Emphasis: New ideas
Glue: Innovation, development
Decision-making style: Collaborative
Decision
Information
gathering
making
Internal sources
Focus on individual
Measure details
External sources
Focus on organization
Model patterns
Culture: Bureaucracy, hierarchal
Physical appearance: Conservative, uniformly similar
Leader style: Organizer, coordinator
Strategic goal: Permanence
Emphasis: Stability
Glue: Rules, policies
Decision-making style: Many approvals, political clout
Culture: Market
Physical appearance: Conservative but individual
Leader style: Hard driver, productive
Strategic goal: Winning
Emphasis: Achievement, accomplishment
Glue: Goal accomplishment
Decision-making style: Rational, fact-based, performance over position
Present-oriented
Short-term
Strives for stability
Ordered by rules

Figure 4.3 A model of organizational culture types.

There are four basic culture types in organizations. In order of their formality and structuring they are:

1. Clan culture
2. Adhocracy culture
3. Market culture
4. Bureaucratic culture

The clan culture is the loosest and least structured. It is encountered in organizations where there is a strong emphasis on individuals and where heroes are common.

Adhocracy culture is also informal and relatively unstructured. It develops in high-tech and scientific development organizations where a great amount of invention or innovation is required. It gets what structure it has from its dedication to development. Cross-functional teams are common.

Market cultures are more structured and formal. They are found in organizations which are under great pressure to produce and where the source of pressure is external. Efficiency is a dominant force.

Bureaucracies are the most formal and structured of the four cultures. They are common in large organizations like federal government and *Fortune* 100 companies. All process is governed by rules, regulations, and protocol. The focus here is also on the individual but, unlike the clan culture, the focus is on the position rather than the person. These are power structures.

The focus on individuals in the clan culture and the bureaucratic culture are derived from their information-gathering behavior. It emphasizes individual contact. Decision-making behavior, on the other hand, tends to separate the cultures into long-term thinkers and short-term thinkers. Long-term thinkers seem to have more tolerance for ambiguity.

Adhocracy and market cultures are more attuned to participation in project teams. Both have goals and emphasis that are consistent with a project team. When a project team is formed, it will exhibit the characteristics of one or the other of these cultures.

In a large manufacturing company, there are pockets of each of these cultures, which leads to fundamental differences of viewpoint on cross-functional teams. These differences of viewpoint are some of the baggage we get when we select team members for their different skills. Good project leadership includes a large dose of recognizing and respecting these differences. They will not change rapidly but, in order to form an effective project team, some of the home organization culture must be compromised for the team culture.

Team building

When we talk about buy-in to project objectives, we are asking the team members to moderate some of their home organization culture with our project culture.* They are being asked to be the representative of their home organization on the project team and to be the representative of the project in their home organization. This is an easier task for the moderately structured cultures, adhocracy, and market than the other two because of their organizational orientation.

When dealing with these different cultures, a good leader phrases his or her desires in each of their terms and checks for verification of understanding. Team members from clans and bureaucracies will have the greatest difficulty in fully joining the team because of their focus on individuals.

Because project leaders cannot depend on having the authority to direct team members, they must develop strategies for influencing others. A single strategy is not effective across all companies, or even across all functional departments, because of the differences in cultures. The cultures of companies and departments are distinctive like the cultures of nationalities. To be effective at influence, a project leader must be sensitive to the local cultures.

Leading the project team is a major responsibility of the project leader. The leader can't stop arguments and disagreements or make people get along together, but the leader must still contribute to unity. The leader is the focus of project team unity. The team has the roots of cohesiveness in its unity of purpose.

The next step for team building is that team members respect the special knowledge and skill of the other members and under-

* Phillips, M.E., Goodman, R.A., and Sackmann, S.A., "Exploring the Complex Culture Milieu of Project Teams," *PMNetwork,* vol. 6, no. 8, November 1992.

stand that those must fit together to complete the job. It is the leader's responsibility to integrate the diverse skills of the team and moderate cultural and personality differences.

There are always professional differences. It is a supreme accomplishment of leadership to use disagreement constructively. For a project team, disagreement is healthy. It contributes other ideas, other approaches, other strategies, and other views. These are desperately needed in the problem-solving process.

The limitation on disagreement is that it should be kept orderly and the emotional content moderated. Otherwise, the discussion will deteriorate into either a shouting match between two competing ideas or into just a chaotic conversation among all.

The purpose of the team is problem solving. This must be the focus of all meetings. The purpose is to solve problems as a group, not to prosecute the guilty or to accuse the innocent. Because projects are a multistage process, there are many handoffs. For example, if A owed B a drawing but was late in delivering it, B is also likely to be late. Too often, we try to make up the lost time by "putting B's feet to the fire" by holding B to its planned delivery date. This is not problem solving. It is further victimizing the victim. Problem solving by the team requires that we accept the fact that if no correction is made, we will be late. Once that happens, we can examine what can be done to help B make up some of the time and where we can make up the balance. However, B is not held accountable for the current delinquency. A witch hunt does not benefit any project.

Accusation destroys teamwork, so the absence of accusation contributes strongly to team building. The project leader will have to reinforce this rule repeatedly. It's worth it—the result will be a stronger team.

The project leader plays a strong role in minimizing external conflicts, too. Because they are goal-oriented, project teams will step on a few toes. This will most often occur as domain infringement. In the public sector, an experienced and knowledgeable public works director told me that his life expectancy in any one government would be six years. He said that if he did anything, he would alienate or upset one-third of the commissioners each year. By the end of six years he would have upset all of the commissioners twice, and they would throw him out. The lesson here is that we cannot eliminate irritating the line organization because we are goal-oriented, but we can try to minimize the irritations. True discussion is one of our best tools for moderating the emotional content in a disagreement.

Appointing the TQM Planning Team

Now that we have full disclosure on the quality level of the company, we can begin to make improvements to the system, right? . . . Wrong! There is still a great deal of preparation remaining before we embark on the improvement journey. Skipping the critical steps subsequent to project execution will result in higher costs, longer schedules, and wasted resources. We now need to appoint the TQM planning team.

There are several aspects of the planning team which we will discuss. If the organization is relatively small, a single planning team will suffice. If the organization is large, however, it may require planning teams at various levels in addition to a quality council. In this case, the quality council identifies the strategic goals but leaves it to the project teams for tactical and operational implementation. There also needs to be a TQM project owner. This is someone who, regardless of the number of project teams, acts as the catalyst for the TQM implementation.

The executive quality council

Much has been written on quality councils, their function and necessity in the scheme of TQM. The beginnings of a quality council occurred when the TQA was reviewed. In most cases, the establishment of a quality council will be required to guide the implementation of TQM. As stated previously, the quality council will double as the planning team if the organization is small. But what is small? In companies with about 100 employees or less, the scope of executive activities allows the president and vice presidents (in total, probably four or less) to maintain a dual role in the process. The guiding, planning, and execution functions are realistic for the executive group in small companies.

The quality council, regardless of function duality, has many important characteristics. First, the council is comprised of select senior executives led by the TQM project sponsor. These are the people who have ultimate responsibility and authority for company policy and administration. As part of this authority, they can provide resources in terms of financial support and people. Capital expenditures for improvements to quality can be provided by the council. Without their commitment, any corporate programs or changes will fade and eventually vanish. The TQM project sponsor must have the council meet on a regular basis to ensure that the implementation occurs as planned and to change direction when necessary. During these meetings they will also review any quality improvement projects. Their responsibility is to approve, modify, or decline nominations for projects. Additionally, the council will review projects in process in terms of schedule, cost, and performance.

The quality council also determines the types of rewards and recognition that will be provided. Rewards may be monetary or in the form of promotions, while recognition might include banquets, ceremonies, and awards.

Top management, via the TQM project sponsor, plays a key role in the success or failure of the TQM implementation. They must be visible and active in the process which is being espoused.

The TQM project owner

Along with the quality council, the selection of a project owner is necessary to implement TQM. The TQM project owner must be someone within the organization. This is very important. A consultant is typically hired because of knowledge and experience, but he or she should not be the TQM project owner. The consultant works closely with the TQM project owner in developing and guiding TQM. Making a consultant the TQM project owner is like shooting the lead buffalo of the herd. When the lead buffalo is gone, the herd is dumbfounded; it will not move. Similarly, when the consultant is gone (after being the TQM project owner), the organization will follow a similar fate. *Note:* Shooting the consultant is not an option.

The characteristics of the TQM project owner or TQM project manager are numerous. First and foremost, he or she must be willing to accept the huge responsibility. The TQM project owner must have a positive attitude toward change, as well. The person selected should be respected in the organization and should maintain a professional demeanor. The TQM project owner may come from any level of the organization, but should have had a reasonable amount of exposure to most functions. Generally, someone in middle-to-upper management is selected for this responsibility. The TQM project owner also has the characteristics described for a project leader.

There are many activities the TQM project owner is required to perform. The following list identifies the key items to be addressed by the TQM project owner.

Select the project leader(s). The TQM project owner must be able to recognize the traits necessary for the project leader. In some cases, a reluctant but qualified individual may have to be sold on the idea of becoming a project leader.

Project leaders/line organizations relationship. The TQM project owner helps to create the matrix organization required for the TQM implementation.

First-level work breakdown. The TQM project owner establishes the work to be accomplished on a macro scale, then assists the project leaders in the remaining work breakdown levels.

Define performance expectations/evaluation. This is the cornerstone of the TQM implementation. Without performance and evaluation methods, top management commitment appears to be weak or nonexistent.

Establish periodic reporting requirements. Reporting requirements may be informal, one-on-one sessions, group meetings, telephone conversations, memos, or reports. The media and frequency is established here.

Represent the project in executive meetings. The TQM project owner is the interface between top management and the TQM project, whether the project is small or large.

Assist in removing impediments to the project. The TQM project owner must respond to obstacles such as departmental barriers, tight budgets, or personnel conflicts, as necessary.

The project leader(s)

The project leader (actual number of leaders is determined by the scope of the project) must meet certain challenges. The TQM implementation environment is dynamic, not static, and the project leader must respond properly to various situations.

Manage tight schedules effectively. Once the project schedule is determined, the project leader is required to "make things happen." Time will slip away quickly if a systematic approach for follow-up is absent.

Work smoothly in an environment of multiple demands. Most likely, everyone will have responsibilities other than the TQM project. This is especially true of the project leader. Demands from the project and line function demands have to be juggled intelligently to prevent a ball from dropping.

Keep work focused on the project goal. Throughout the life cycle of the project, factors intercede which attempt to divert or derail the project. If one objective is to train a particular department in SPC, there may be a manager who would like to have a different type of training done. It is the responsibility of the project leader to remain focused on the original objective.

Maintain cohesiveness on the project team. The project leader must develop the team to work on the project goal collectively, not individually.

Use professional tensions constructively. This means using disagreements—and there will be disagreements—in a positive manner. Disagreements bring about new ideas and viewpoints, which can lead to creative solutions. The trick is not crossing the fine line from order to chaos.

Maintain a problem-solving attitude under pressure. There will be pressure during the course of the project. Pressure generally excites people, usually in a negative sense. Controlling the situation by staying composed will serve to solve the problem.

Minimize external conflict. Although the conflicts cannot be eliminated, they should be kept to a minimum. Cooperation and compromise with outside influences will help achieve the project goal.

The project team(s)

The project team (again, actual number of teams to be determined by the scope of the project) consists of those people who are knowledgeable and who are part of the process being addressed. A common error is to exclude someone who is part

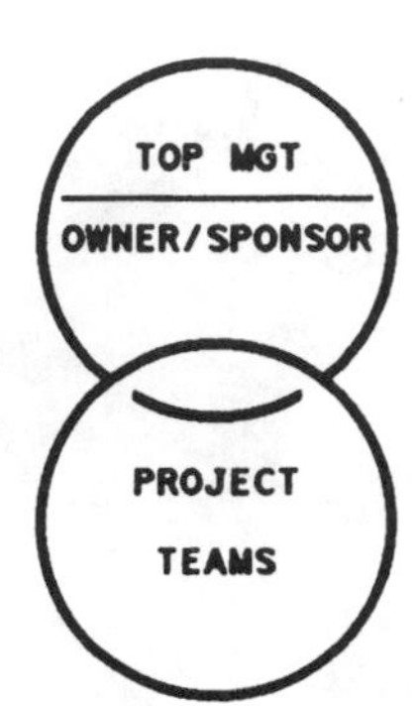

Figure 4.4 TQM relationships in a small organization.

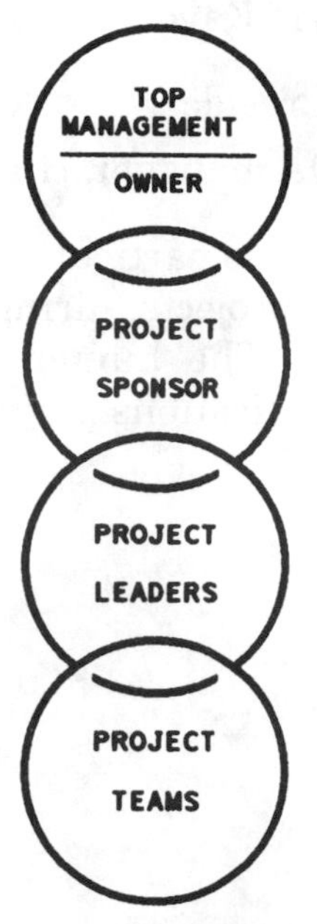

Figure 4.5 TQM relationships in a large organization.

of a process. When a project is complete and an individual becomes aware of the activity, there are two possible responses. A more efficient method may be feasible or the person may become an obstacle or saboteur. Either way, the project may have to be resurrected.

There may be training required for the project team. Problem-solving skills and group dynamics might be necessary for some or all of the team members. This will be especially true for organizations where individual accomplishments are emphasized over the team approach.

Figure 4.4 illustrates the relationships between the quality council, TQM project owner, and project leaders/team in a small company. Figure 4.5 illustrates the same relationships in a large company.

Our TQM case study organization consists of a project sponsor, a project owner, and two project leaders. They are:

Project owner: C. P. Kaye

Project sponsor: President

Project leaders: QA manager, Human resources (HR) manager

The QA manager in this particular implementation will be leading two of the three projects: hiring the consultant and statistical implementation. The human resources manager will be directing the communications project.

Chapter

5

Step 3: The Work Breakdown Structure

The Work Breakdown Structure (WBS) has been given a minor role in project management literature for several years. This is unfortunate because it is our primary analytical tool. Without a clear approach to developing a WBS, we deprive ourselves of a thorough analysis of the work which needs to be performed to accomplish the project objectives and satisfy the project purpose. Further, the WBS provides the coordinating framework which is necessary to integrate all the different forms of information in a project. It also provides definition of the different levels of data summary for reporting.

With the popular spread of project management software, there has been an emphasis on the modeling capability of the software. The capability of most software to enforce or even encourage a WBS analysis is rare. It is assumed away as a minor phase in the development of a logic network of the sequence of tasks. The network model may imply the presence of a WBS by having an intelligence-bearing coding scheme for the tasks in the network.

Background

In the 1960s, when we were drawing logic networks by hand, we spent more time on developing the WBS. It provided us with a clear picture of the work to be performed, the resources needed, and a framework for collecting costs. When it was completed, it gave us an almost one-to-one translation from the tasks of the WBS to the nodes or arrows of the logic network diagram. Once work started, we could use the structure of the WBS to accumulate cost data and summarize status for management reporting. The logic network was used only for scheduling.

The structure of a WBS comes from the engineering roots of project management. It has a structure similar to that of a bill of material. Design engineers were accustomed to analyzing the parts of a new product with a bill of material. This structure started with the whole, and began desegregating it into smaller and smaller pieces until parts which were practically indivisible were identified. For example, an automobile starts with a designer's drawing. It is then divided into major systems: the body, power train, suspension, electrical systems, etc. Each of these systems is divided into assemblies, assemblies into components, and components into parts. For example, an instrument panel is divided into its assemblies (instrument cluster, heater ducts, etc.); the instrument cluster is divided into components (gauges, dials, etc.); gauges are divided into parts (meter, connector, screw, etc.).

The purpose of the bill of material and the WBS is the same: to facilitate the building of a system. The bill of material helps build an automobile. The WBS helps build the program plan and its logic model of the flow of work. Each of these end products can be built without the use of its analytical tool, but the results are inconsistent. That is, the structure of the result would depend on the whims of the person guiding its development.

Variations in Work Breakdown Structure

We encounter a wide variety of WBS structures as we work with professionals from different lines of work, such as design engineers, information systems personnel, interior design people, and development scientists. Each has his or her own view of the steps required to complete a job. However, the underlying flow of work seems to remain constant.

This variation seems to come from the perception of the originator of how his or her work is structured. The general tendency is for each of us to think that the whole world is structured according to how we see our part of it. Consequently, a systems engineer sees an automobile development as a collection of systems. A finance officer may see an automobile development as a collection of cost centers. A functional manager may see an automobile development as a collection of skills.

Conceptualizing a Work Breakdown Structure

The problem facing a project developer is which of the possible structures should be used for a particular project's WBS. This is perhaps the most intellectually challenging problem of project management. It might seem that the choice makes no difference. However, when carried out, each of these different patterns makes a specific type of information more accessible than the others, and some information may even become inaccessible. For example, a cost-center-based WBS will lead to a project in which the emphasis is on cost data; a skills-based WBS will lead to a project in which the emphasis is on people and resource utilization; a phased structure will lead to a project in which the emphasis is on the flow of work. The point is that the structure of the WBS influences the emphasis of the project. Herein lies the answer to the problem of which structure to use.

There are three basic structures for WBSs*:

- Product structure
- Process structure
- Organizational structure

1. *Product structure* breaks the work into packages similar to that of a bill of material. It will start with the final deliverable item. This is broken down into its major assemblies and the information and services which are to be delivered with the product. Assemblies are broken down into subassemblies and the work necessary to test and assemble them. This breakdown

* Youker, Robert, "A New Look at Work Breakdown Structure," *Proceedings of the Project Management Institute Annual Seminar/Symposium,* Calgary, Alberta, Canada, 1990.

continues down into the parts. This is a good structure for hardware items. It places emphasis on quality. Organizational responsibility for work and the time phasing of the work are implicit in this structure.

2. *Process structure* breaks the work into time phases. All work on assemblies, subassemblies, components, and parts is listed in each phase. This structure highlights where in the process the work is taking place. It is a good structure for emphasis on schedule. The structure of the product and organizational responsibility are implicit in this structure.

3. *Organizational structure* starts with the project owner's organization, then breaks it down into the teams which must perform the work. This structure starts off looking like an organization chart, but differs in that it focuses on the work to be performed in each organization. By doing so, it focuses on inputs to the project from each organization. The flow of the process and the development of the final product are implicit in this structure. This is a good structure for organizational changes.

In practice, the different structures are not mutually exclusive. Frequently, functional organizations or subproject teams are organized around the major assemblies of the product. For example, in automobile development, one division or team may be responsible for development of the body, another for the power train, and still another for the interior. (See Fig. 5.1.) Many information systems organizations are structured by the phases of software development (such as systems analysis, systems design, systems development). (See Fig. 5.2.) This structuring of organizations reflects the emphasis of the organization. A WBS, therefore, can be structured organizationally and also incorporate one of the other structures automatically. We must take note of this phenomenon when we plan a WBS.

Pick the structure which places the emphasis on the primary management goal of the project, whether it is high quality, quick development, cost control, allocation of scarce resources, etc. All of these are noble management goals, but only one can be primary. That decision must be made in the definition of the project prior to the development of the work breakdown structure.

Once the decision has been made about the primary management focus of the project, the WBS can be structured. This decision manifests itself in the first level of breakdown. This level is

WBS Structure

Level	Description	Code	
Level 1:	Functional group		
	1. Product development	E.	
	2. Program management	M.	
Level 2:	Product engineering office (PEO)		
	(Product development)		
	Vehicle engineering	VE.	
	FMVSS test and analysis	TA.	
	PTPD	PP.	
	Body engineering	BE.	
	Climate control	CC.	
	Powertrain systems	PT.	
	LTE chassis and P/T design	TE.	
	LT CAD	LC.	
	Vehicle build	VB.	
	Transmission and clutch	TC.	
	Engine fuel handling	EF.	
	Electronics	EL.	
	(Program management)		
	Program planning and control	VO.	
	Financial control	FI.	
	Vehicle development/durability	VD.	
	NAD	NA.	
	Manufacturing operations	MF.	
	Purchasing	NP.	
	Marketing	MK.	
	Customer evaluation	CU.	
	LT safety	LS.	
	TOCO	TO.	
	ASO	AS.	
	LT timing	LT.	
Level 3:	PMT or deliverable	00.	
Level 4:	Phase		
	Program definition	⟨PD⟩	01.
	Program implementation	⟨PI⟩	02.
	Theme decision	⟨TD⟩	03.
	Program confirmation	⟨PC⟩	04.
	Prototype readiness	⟨PR⟩	05.
	Sign-off	⟨SO⟩	06.
	Launch readiness	⟨LR⟩	07.
	Final program status	⟨J1⟩	08.
Level 5:	Activity	00.	
Level 6:	Task	00.	

Sample: E.BE.02.03.04.00

E. = Product development
BE. = Body engineering
02. = Body shell
03. = Theme decision phase ⟨TD⟩
04. = SP design freeze
00. = No task presently assigned

Figure 5.1 Work Breakdown Structure codes for a vehicle development.

Checklist Work Breakdown Structure

Systems analysis
- 10X Organization
- 11X Business strategy
- 12X Information needs
- 13X Present status
- 14X Application strategy
- 15X Hardware and software strategy
- 16X Organization strategy
- 17X Implementation strategy
- 18X Information plan
- 19X Project initiation
- 1AX Project definition and planning
- 1BX Approve and implement plan

Systems design
- 20X Organization
- 21X Hardware and systems software direction
- 22X Application software evaluation and design
- 23X Functional and technical specifications
- 24X Installation schedule
- 25X Cost/benefit analysis
- 26X Hardware and software selection
- 27X Management review and approval

Systems development
- 30X Organization
- 31X Detailed design
- 32X Systems software development
- 33X Hardware and software installation
- 34X User procedures development
- 35X Conversion preparation
- 36X Programming
- 37X Systems test
- 38X Conversion
- 39X Postconversion review

Maintenance and enhancement
- 40X User liaison
- 41X Performance tracking
- 42X Information plan coordination
- 43X Request classification
- 44X Priority setting
- 45X Change analysis
- 46X System modification
- 47X Implementation
- 48X Status evaluation

Figure 5.2 Work Breakdown Structure codes for software development.

frequently divided into phases in automotive projects because flow of work is critical to the acceleration of product development. If the primary management focus is different, then the labels applied to the first level of breakdown should be appropriate (for example, major assemblies, deliverables, cost centers, divisions, major functions).

Structure of the Work Breakdown Structure

There has been a great deal of confusion regarding the labels applied to the levels in a work breakdown structure. To be consistent, I suggest that we use the classical naming convention, which is:

Top Level (level 0)	Project
First Level (level 1)	Phase, deliverable, etc.
Second Level (level 2)	Activity
Third Level (level 3)	Task
Fourth Level (level 4)	Procedural step

Figure 5.3 illustrates this naming convention. Training materials and other project information should reflect this convention as a coordinating device.

The classic naming convention for the levels of a project WBS come from the early days of project management. That convention has been adapted to projects of the size which automotive and other manufacturing firms usually have.

This WBS naming convention was lost, thanks mostly to the software publishers. As a group, they have a tendency to use "activity" and "task" as interchangeable terms for the work packages which we schedule and to which resources are allocated in a network. This is an error in consistency and causes confusion. We should do what we can to reduce this confusion.

In the classic nomenclature, the task level is always the one which we schedule and for which we allocate resources. This can also introduce some confusion, especially when we have a WBS of only one or two apparent levels. The work packages which we schedule are still the tasks. We drop the unused levels between the project at level 0 and the task level, as illustrated in Fig. 5.4.

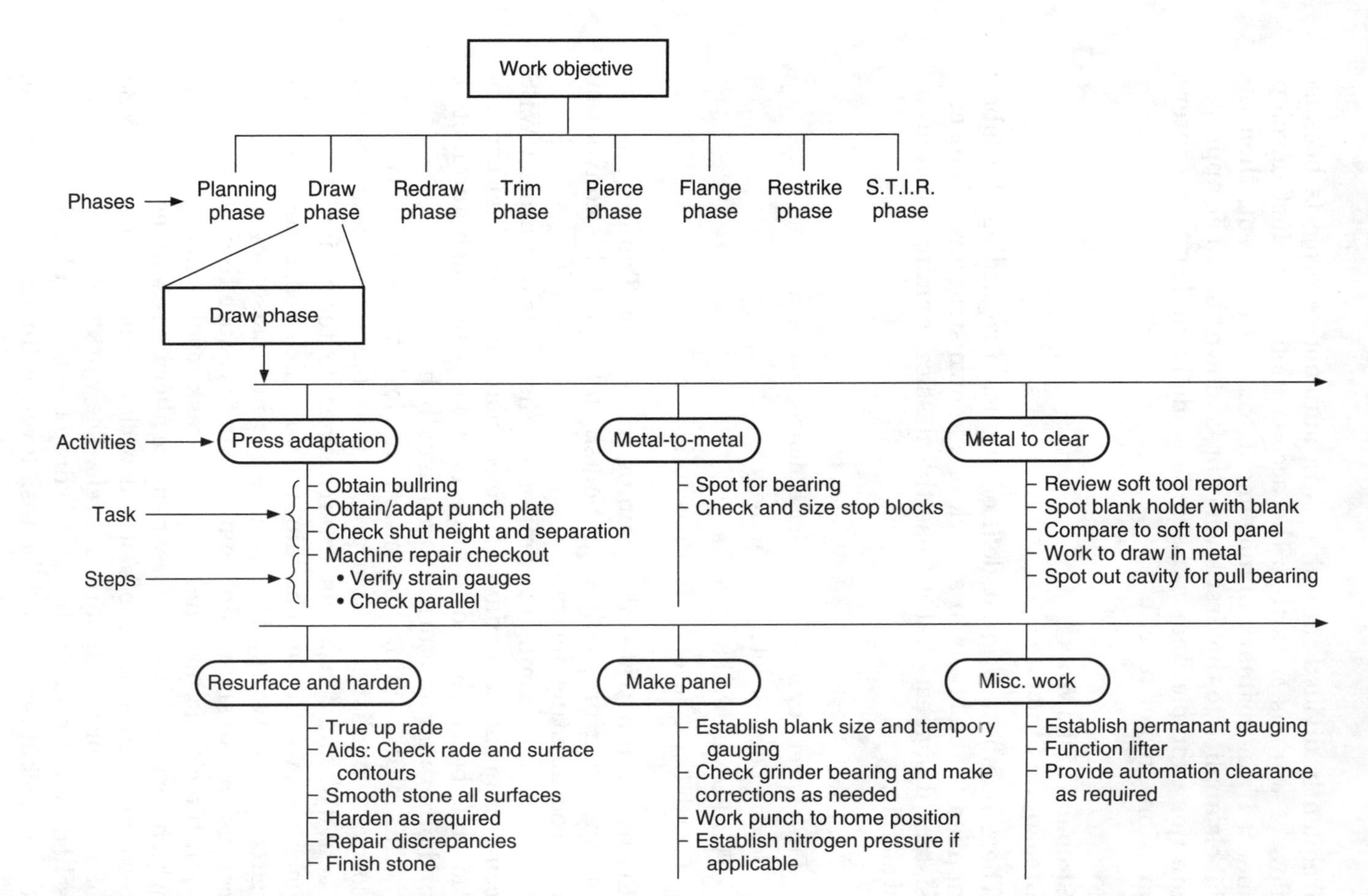

Figure 5.3 Work Breakdown Structure for metal stamping tryout.

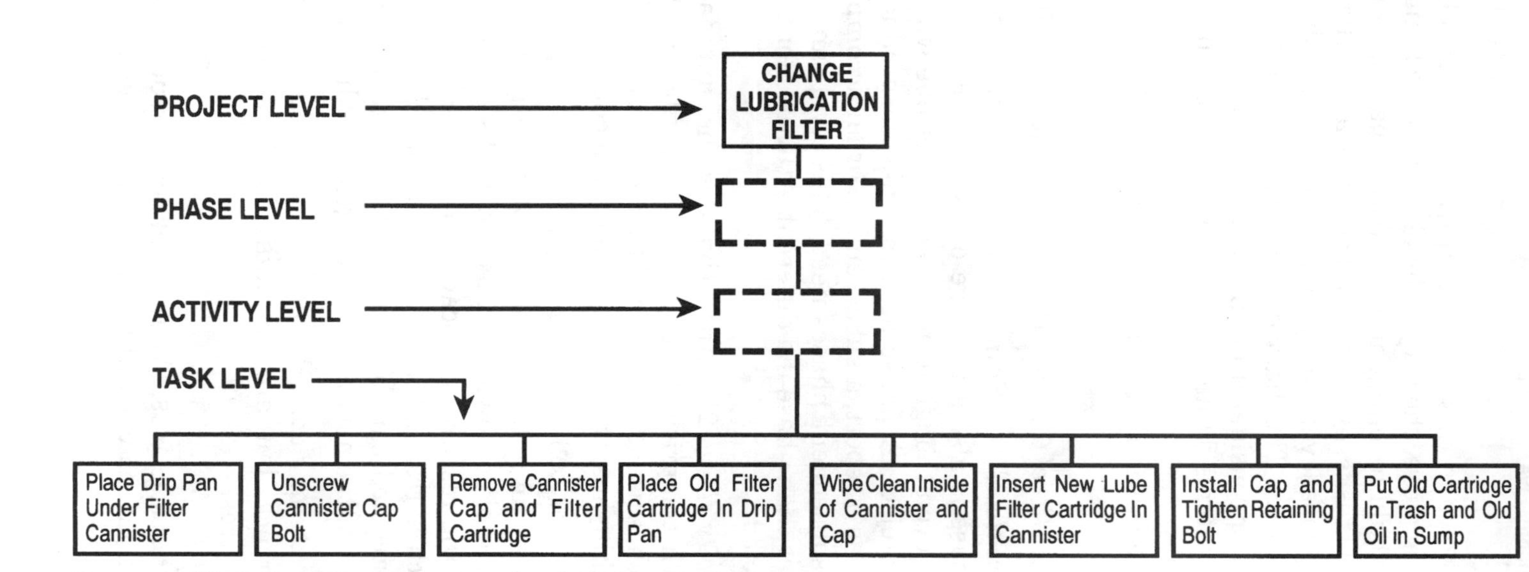

Figure 5.4 Work Breakdown Structure for changing a lube filter.

As shown in this example, if we were to develop the WBS for changing a tire, the project level would be "changing a tire," and there would be only one row of work packages under it. These would be tasks, even though they appear at what usually would be the phase level. We have just omitted the phase level and the activity level. The diagram illustrates this configuration.

Another consideration in the development of a WBS is that all primary items do not have to be broken down to the same level of detail.

The definition for each of these levels for a process structure is as follows:

Project (level 0). Summary of all phases of the project.

Phase (level 1). Summary of the activities which comprise it. A phase is also a package of work which produces a deliverable, even though it may be a deliverable which is internal to the project. A phase always makes a product of some type, such as a report, a subassembly, or an accomplishment which the subsequent phases need. This level can have a label different than phase if a different structure for the breakdown is selected.

Activity (level 2). An activity is the summary of the tasks which comprise it. An activity describes a package of work which is usually performed by a section or work group, such as machine maintenance, simulation modeling, market analysis.

Task (level 3). A task describes the work which would be done by an individual or a small work group, and is not going to be broken down any further. Tasks are the work scheduled and the resources allocated to them. They are also the work which is translated into the network diagram. If the WBS is constructed carefully, translation of tasks from the WBS to nodes of the network logic diagram is almost one for one.

Procedural step. Procedural steps of a task can be listed for reference, but are not individually scheduled or allocated resources.

Outline format for WBS. Work breakdown structures were originally hand drawn as vertical tree diagrams. They started small at the top with the project, then spread out as increasingly more detail work was added at successively lower levels. Today, an indented outline form is used because it is much easier to work with than a piece of paper hung on 30 feet of wall somewhere. The information is the same in both formats. In the outline format, we use indention to signify levels of the WBS.

Recently, some of the software publishers have given us an additional incentive for the outline format. We can visualize the WBS and its codes at the left of a spreadsheet. To the right of each task, additional information can be added, such as task duration, resource needs, and cost. So, the outline form becomes the base of a project worksheet. Estimating of this project data will be dealt with in Chap. 6.

WBS codes. Work breakdown structure codes are sometimes confusing when first encountered. They may seem like some kind of mysterious code hung on each item in the WBS.

The codes in WBSs are the feature which allows a WBS to coordinate information on work progress, problems and issues, resource utilization, and cost. It is an intelligent system of letters and numbers which identifies to which organization, assembly, and phase each task belongs. Each task's code is unique to it. When a task code is associated with a problem or change order, it tells us where in the project plan it is occurring. We can then assess schedule, resource, and cost impact.

Work breakdown structure codes are designed as position-significant strings of letters and numbers as shown in Figs. 5.1 and 5.5. For ease of reading, the positions are usually separated by a dot. (See Fig. 5.6.) The scheme is as follows:

Position 1	Category of the project
Position 2	Project identifier
Position 3	Level 1 identifiers: phase, assembly, or functional organization unit
Position 4	Level 2 (activity) identifier
Position 5	Level 3 (task) identifier, assigned in sequence if it is the lowest level of the WBS (this sequence should skip two to five numbers or letters to allow for insertion of overlooked tasks)

Project:	Upgrade Temper mill	Expected start date:			2/15/89
Leader:	D. W. Witalas	Expected completion date:			12/31/89
		Requested completion date:			12/31/89

WBS code PH ACT TSK	Description PH ACT TASK	Responsible resource	Work days	Elapsed days	Predecessors
1 0 0 0	Organize and plan				
1 1 0 0	Form project team	ENGR			
1 1 1 0	Assign IE	IELT	?	?	
1 1 2 0	Meet task force members	DWW	1.0	10.0	
1 1 3 0	Work breakdown structure	DWW	0.5	5.0	
1 2 0 0	Review previous work				
1 2 1 0	Identify previous IE work	DWW	1.0	5.0	
1 2 2 0	Read through previous work	DWW	2.0	2.0	
1 3 0 0	Identify potential benefits				
1 3 1 0	Marketing and sales wish list				
1 3 1 1	Identify and contact key marketing and sales personnel	DWW	0.5	4.0	
1 3 1 2	Summarize feedback	DWW	1.0	20.0	
1 3 2 0	Engineering wish list	ENGR	?	?	
1 3 3 0	Operator's wish list	OPER	?	?	
1 4 0 0	Review potential benefits with eng, sales, and operators	TEAM	?	?	
2 0 0 0	Establish current base				
2 1 0 0	Investigate database				
2 1 1 0	Mainframe	DWW	2.0	5.0	
2 1 2 0	PC	DWW	2.0	5.0	
2 1 3 0	Other	DWW	2.0	5.0	
2 2 0 0	Analyze current operations				
2 2 1 0	Time study operations	OPNS ANAL	2.5	2.5	
2 2 1 1	Recap time study	DWW	3.0	4.0	
2 2 1 2	Publish results	DWW	2.0	15.0	
2 2 2 0	Yield and rejects	DWW	2.0	5.0	
2 2 3 0	Productivity	DWW	2.0	5.0	
2 2 3 1	#28 temper				
2 2 3 2	Other related	DWW	1.0	4.0	
2 2 4 0	Quality	OPNS	?	?	
2 2 5 0	Increased realization				
2 2 5 1	TMW	DWW	1.5	4.0	
2 2 5 2	Higher-grade product	DWW	1.5	4.0	
3 0 0 0	Define state of the art				
3 1 0 0	Visit other mills	ENGR	?	?	
3 1 2 0	Contact mill builders for proposals	ENGR	?	?	
4 0 0 0	Define future state				
4 1 0 0	Mill characteristics				
4 1 1 0	Scope	ENGR	?	?	
4 1 2 0	Cost	ENGR	?	?	
4 2 0 0	Tangible benefits				
4 2 1 0	Yield and rejects	DWW	2.0	5.0	
4 2 2 0	Productivity				
4 2 2 1	#28 temper	DWW	2.0	5.0	
4 2 2 2	Other related	DWW	2.0	5.0	
4 2 3 0	Quality				
4 2 3 1	Customer needs/changing market	SALES	?	?	
4 2 3 2	Risks of not upgrading	SALES	?	?	
4 2 4 0	Increased realization				
4 2 4 1	TMW	DWW	3.0	5.0	
4 2 4 2	Higher-grade product	DWW	2.0	5.0	
4 2 5 0	Intangibles	TEAM	?	?	
5 0 0 0	Write and present proposal				
5 1 0 0	Final review and proposal design				
5 1 1 0	Establish proposal objectives	TEAM	?	?	
5 1 2 0	Develop proposal strategy	TEAM	?	?	
5 2 0 0	Write proposal	TEAM	?	?	
5 3 0 0	Present proposal	TEAM	?	?	

Figure 5.5 Work Breakdown Structure and planning worksheet.

Program Office
Work Breakdown Structure

09-Jan-92

WBS code	Description	Responsible	Resource	Remaining duration from 10/1/91	Predecessors
1.	Design release				
1.1.	EP design freeze				
1.1.1.	Chief eng's design aid buck sign-off	G. Langner	3 package eng	55	4.1.6.
1.1.2.	No adjust build (NAB) locator scheme	T. Sweder	3 body eng (350 DA)	Compl	
1.1.3.	Verify CDVT's inspection standards		3 B&AO	5	
1.1.4.	EP part release		16 PEO release	50	1.1.5.
1.1.5.	Prepare drawings for EP review	F. LaPlaca	2 veh prog control	5	
1.2.	VP design freeze				
1.2.1.	Prepare drawings for VP review	F. LaPlaca	2 veh prog control	5	4.1.15.
2.	Material management				
2.1.	Supplier interface				
2.1.1.	Supplier sourcing	F. Parfen	5 NAAO purch (260)	Compl	
2.1.2.	Supplier visits				
2.1.2.1.	Three months before MRD		? WIN88 eng	?	2.2.2.,

Figure 5.6 Work Breakdown Structure with data columns.

There may be as many additional levels of breakdown as thought necessary. But the number of levels may be limited by the space in the field of the software. Usually this is 10 to 16 spaces. The entire code for each task must fit into this space. This includes the dots as well as the identifiers. Identifiers are usually two numbers or characters, as shown in the examples.

A description of each task should be associated with each task code. This should describe the work to be done, what starts it, and what signals its completion.

Level of detail. A recurring question when developing a WBS is the level of detail to which it should be carried. Several criteria can be applied for making this decision. In general, more detail makes more work in tracking progress. On the other hand, less detail may fail to track certain items, some of which may be important.

The following is a set of guidelines for establishing a comfortable level of detail in your project:

- The detail level must contain tasks which have quantifiable inputs, outputs, and identifiable points of progress.
- The duration for any one task should not be more than 5 percent of the total project duration. All durations should be short enough to make the task manageable.
- A risk is involved with large duration tasks when progress is assessed. If the duration is 100 days and the progressing error is plus or minus 10 percent, the estimated dates could be miscalculated by plus or minus 10 days.
- The length of the update cycle will also determine the length of detail tasks. Updating 150 one-day tasks on a daily basis creates a large administrative burden. Conversely, updating a series of five-day tasks once every six months does not provide much accuracy in the update or schedule projections. Generally, no task duration should be longer than three update cycles.
- Each task should have one person accountable for all handoffs and completion.
- The level of detail should be such that resources and costs can be managed, as well as schedule.

The TQM Work Breakdown Structure

Since we have identified the strengths and weaknesses of the company (using the TQA) and we have selected the planning team, the next step is to define the work with a Work Breakdown Structure (WBS). Previously, we learned how to develop a WBS based on the results of the TQA and the direction of top management. The concept of the WBS will now be applied to the TQM implementation.

Because of the potential complexity and cross-functional nature of the TQM implementation, the WBS form most appropriate is the product structure. Consequently, the primary management goal of the TQM implementation should always be improving quality. Remember the definition of quality and the Quality Matrix™. Productivity and profitability will naturally increase as a result of the improvements to quality, and therefore should not be stated as primary management goals.

From Chap. 4 we found our case study company to have the following deficiencies:

1 point	Internal communication needs improvement; no formal system in place
0 points	No use of statistical methods; system not evident on shop floor

You can see that one of the items, communications, is from the management system and the other, statistical methods, is from the technical system. If we assume that these are the only areas needing improvement,* the WBS can be created in the following manner. The project, or level 0, is the "TQM implementation." The two-phase levels of our example are "Establish communication system" and "Implement statistical methods."

To improve the flow and understanding of the following information, these abbreviations are placed by the WBS code:

PR	Project
P	Phase
A	Activity
T	Task

* It is very unlikely that a company would be deficient in only these two areas.

Each part of the WBS (in bold) is followed by a brief explanation. The work breakdown structure in Fig. 5.7 illustrates the final product of our example.

0.0.0 (PR) TQM implementation

1.0.0 (P) Establish communication system

In evaluating the entire communications system of the company, external communication appears to be acceptable. Internal communication, however, is deficient. Suppliers and customers have a clear understanding of the company's policies and procedures, and there is a system for ongoing written and verbal communication. The survey of suppliers and vendors revealed some minor, nonrecurring problems, but the overall rating from both was good.

Internal communication needs a great deal of improvement. Many employees felt they had no idea what the company direction was, who the customers were, and what the results of audits and visits were. Some functions held communication meetings on an irregular basis, while others did not hold any meetings. The subjects tended to be inconsistent as well.

2.0.0 (P) Implement statistical methods

Data is being collected on the various parts as they are produced, and it is being recorded. The information remains in the form of raw data, filed the day after production with little or no review.

3.0.0 (P) Hire consultant

Before the other two phases can begin, the consultant must be hired. The skills of the consultant are needed for both the communications phase and the statistical methods phase.

As we begin to build our WBS, it is apparent that phase 1.0.0 cuts across all functions, while phase 2.0.0 is confined to the manufacturing area. Because of this, the level 2 activities for phase 1.0.0 will not identify any particular area. The activities for phase 1.0.0 are:

1.1.0 (A) Survey employees

The initial survey in the TQA identified the fact that there was a problem with internal communication. This survey will help

0.0.0 Project: TQM implementation
Project owner: C. P. Kaye

Expected start: 1/1/93
Expected completion: 12/31/93
Requested completion: 12/31/93

Code (Phase Activity Task)	Description (Phase Activity Task)	Responsible resource	Work days	Elapsed days	Predecessors
1.0.0	Establish communication system				
1.1.0	Survey employees				
1.1.1	Write survey				
1.1.2	Survey review/approval				
1.1.3	Identify employees				
1.1.4	Schedule survey				
1.1.5	Print survey				
1.1.6	Conduct survey				
1.2.0	Determine communication media				
1.2.1	Tabulate/review results				
1.2.2	Approve media				
1.3.0	Set schedule				
1.3.1	Identify resources				
1.3.2	Publish calendar				
2.0.0	Implement statistical methods				
2.1.0	Identify appropriate techniques				
2.1.1	Hire consultant				
2.1.2	Review department operations				
2.1.3	Hold review meeting				
2.2.0	Organize training				
2.2.1	Supervisor subjects				
2.2.2	Operator subjects				
2.2.3	Determine class no./size				
2.2.4	Schedule classes				
2.2.5	Training materials				
2.2.6	Conduct classes				
2.3.0	Evaluate effectiveness				
2.3.1	Discuss with supervisors				
2.3.2	Discuss with operators				
2.3.3	Review charts				
3.0.0	Hire consultant				
3.1.0	Consultant search				
3.1.1	Identify consultants				
3.1.2	Telephone screening				
3.1.3	Preliminary selection				
3.2.0	Conduct interviews				
3.2.1	Set appointments				
3.2.2	Interviews				
3.3.0	Select consultant				
3.3.1	Review information				
3.3.2	Group meeting				

Figure 5.7 TQM WBS worksheet.

determine to what type of communication the employees would be most receptive.

1.2.0 (A) Determine media

Review the survey results and decide on the best form of communication, for both the company and the employees.

1.3.0 (A) Set schedule

Whether it is determined that communications meetings, a newsletter, or some other media is the best means, establish a schedulc for the particular type of communication.

1.4.0 (A) Create system

The new system, whatever it is, will have to be planned and implemented. A newsletter, for example, will require a format, materials, printing, subjects, authors, etc., to be successful.

The activities for phase 2.0.0, "Implement statistical methods," are:

2.1.0 (A) Identify appropriate techniques

Since no SPC is in existence, an evaluation of the operation needs to occur. Different processes require different control methods, so selection is crucial at this point.

2.2.0 (A) Organize training

We need to determine the specifics of the training. This includes who will be trained, class size, location, materials, etc.

2.3.0 (A) Evaluate training effectiveness

In order to determine the level of comprehension, there must be an evaluation method. Along with questions for the participants at the end of each workshop, discussing the information one on one and reviewing the actual charts should be done.

The activities in phase 3.0.0, "Hire consultant," are:

3.1.0 (A) Consultant search

A certain amount searching will be required to find candidates to help with the implementation. The consultant should have experience with surveys and SPC.

3.2.0 (A) Conduct interviews

Each candidate for the consulting position should be treated as if the company were hiring a full-time employee. Personality as well as skills must be taken into consideration.

3.3.0 (A) Select consultant

A review of the best-qualified candidates is required. The final selection is made in this activity.

We are ready to break down each activity into its tasks. The tasks for each activity in phase 1.0.0, "Establish communication system," would be:

1.1.0 (A) Survey employees

1.1.1 (T) Write survey

This task would consist of the survey development, including the response type, number of questions, outside assistance, etc.

1.1.2 (T) Survey review/approval

The survey will have to be reviewed for accuracy, content, grammar, etc. Modifications, additions, or deletions are made and the approval is given to proceed.

1.1.3 (T) Identify employees/departments

A complete list, by department or function, of company employees must be procured to establish the number of surveys required and the location breakdown.

1.1.4 (T) Schedule survey

A schedule is developed based on the availability of the employees in various departments. Vacation, sick leave, disability, and other absences need to be determined. At least one makeup session will have to be scheduled to allow for employees who are not able to take the survey when scheduled.

1.1.5 (T) Print survey

The printing method will have to be determined. It may be done internally via a photocopier or it may have to be sent to a printer.

1.1.6 (T) Conduct survey

The schedule will be followed as planned (hopefully) in item 1.1.4.

1.2.0 (A) Determine communication media

1.2.1 (T) Tabulate/review survey results

The results of the survey will have to be summarized in some manner, and a meeting will be held to review the summary. A recommendation is made on the best media.

1.2.2 (T) Approve communication media

Because there will be further commitment of resources, an upper-level approval will most likely be necessary.

1.3.0 (A) Set schedule

1.3.1 (T) Identify the resources

There must be an evaluation of the resources required in order to set the schedule. The availability of personnel, funds, equipment, etc., will dictate the frequency of the communications and, subsequently, the schedule.

1.3.2 (T) Publish calendar

When the resources have been identified, a schedule or calendar can be published to inform the employees of when the communication will occur. Survey results and method of communication may also be included.

Phase 2.0.0 activities are also broken down into the individual tasks.

2.1.0 (A) Identify appropriate techniques

2.1.1 (T) Review department operations

There may be more than one department which is in need of statistical methods. As mentioned previously, there may be several different techniques which will be applied in the various processes.

2.1.2 (T) Review meeting

This meeting is designed to inform the various managers of the results from task 2.1.2. Any questions about SPC are fielded here, and a general "buy-in" should occur.

2.2.0 (A) Organize training

2.2.1 (T) Supervisor subjects

Generally, the supervisors will be trained at a slightly higher level of detail than the operators. The topics are identified here.

2.2.2 (T) Operator subjects

The topics for the operators are determined.

2.2.3 (T) Determine class number/size

There will be a certain number of classes held to accomplish all the required training, and the class size will be optimally 7 to 10 people. Classes with specific participants are identified here.

2.2.4 (T) Schedule classes

As with other scheduling requirements involving groups of people, vacations, holidays, sickness, etc., will have to be considered.

2.2.5 (T) Training materials

Training materials will have to be developed or purchased. Other items, such as calculators, rulers, and scratch paper, must also be included. Decisions may need to be made on larger expenditures, such as an overhead project if required.

2.2.6 (T) Conduct classes

The schedule, as established in item 2.2.4, will be followed.

2.3.0 (A) Evaluate effectiveness

2.3.1 (T) Discuss with supervisors

Questions will be asked of the supervisors to determine their level of knowledge after the training is complete. These discussions are particularly important because the chart review (item 2.3.3) deals directly with operator knowledge.

2.3.2 (T) Discuss with operators

Questions will be asked of the operators to help determine their level of knowledge.

2.3.3 (T) Review charts

A thorough review of the charts/statistics being generated is also required to determine the training effectiveness. Items on the charts will probably generate questions for the discussion with the operators (item 2.3.2).

3.1.0 (A) Consultant search

3.1.1 (T) Identify consultants

This is accomplished through advertising, looking through trade magazines, or word of mouth. In some cases, there may be a consultant who was used previously on a quality-related subject.

3.1.2 (T) Telephone screening

A large amount of information can be obtained through a telephone screening. This will eliminate those who are too expensive, do not have the appropriate skills, or have a personality not congruent to the corporate culture.

3.1.3 (T) Preliminary selection

A "first pass" review of all the consultants screened by telephone will yield a core group to further evaluate.

3.2.0 (A) Conduct interviews

3.2.1 (T) Set appointments

Appointments have to be scheduled, taking into consideration the availability of all those involved in the interview process.

3.2.2 (T) Interviews

The actual interviews are held for completion of this task.

3.3.0 (A) Select consultant

3.3.1 (T) Review information

A final review of all the information on each consultant is done by the individuals involved.

3.3.2 (T) Group meeting

A group meeting is held to discuss and finalize the selection of the consultant. At this point, the decision is made and the consultant is notified.

There may be other activities and tasks which could be inserted in the WBS presented here. Each case is different and will need a certain degree of customization.

Chapter

6

Step 4: Estimate Task Durations, Resources, and Costs

Background

In some industries and some companies, estimating the time, resources, and cost is a common practice, and in construction it is developed to the level of a profession. In manufacturing, companies doing job-shop work must be skilled at estimating or they perish. In mass production or stream production companies, estimating is seldom a familiar practice among engineers and is relegated to a cost-estimating department in finance. Consequently, this chapter will be rather basic to some readers, while others will find it contains sensible but new ideas. The information presented here will be the fundamentals.

If a total cost for the project were to be estimated in the definition of step 1, the estimate arrived at in this step would be much more accurate. We are estimating the cost of each task. The total cost of the project will be the aggregate of these task detail estimates. Estimates which are built from the detail upward tend to be quite accurate if nothing is overlooked. In project management, we avoid this problem by using the tasks of the work breakdown structure as our checklist.

The form shown in Fig. 6.1 is the vehicle for developing our estimates for the tasks. It starts with the WBS task number on the left, then works to the right through columns which provide useful information. This format is similar to some of the soft-

PMT: ________
Leader: ________
Date: ________

Item no.	Description	Responsible	Resource Qty.	Duration	Input needed	From whom

Figure 6.1 Work Breakdown Structure worksheet.

ware formats for entry of task data. If a software package is being used which has this format, it can be substituted for this format, providing the people doing the estimating have access to the computer.

An explanation of the data requested in the columns follows.

Item number. Item number is the column in which to enter the WBS code for each task being estimated. In practice, the WBS tasks will be selected so that only the ones relevant to a team or organizational unit will be entered on the form. However, a copy of all the tasks and their codes need to be used to code the last two columns which identify the inputs which drive the task.

Description. This is the task name from the WBS. It should match so that cross-references are clear.

Responsible. This can be a manager's name or the name of his or her unit. This is the person under whom that task's work will be performed. It locates the work in the organization for us.

Resource and quantity. What types of people and how many will it take to perform this task? Identify these people by profession, skill, or craft. This information frequently interacts with the task duration in the next column. In tasks where there are several different types of resources needed, list each one on a separate line under the task so their work durations can be identified. For example, installing a machine may require a crane for an hour, even though the total task duration is three days. The crane and operator each need to be listed for an hour on the day needed. Other resources, like millwrights, may be spread across the duration of the task. Fractional durations are permitted for skills like supervision.

Duration. This is the column for the duration of the task or the duration of the use of a resource within the task duration. The most common unit for duration estimates is days. Short projects of not more than a couple of weeks are frequently scheduled in hours. Long projects of several years are scheduled in weeks. The units in which duration is estimated interacts with the progress-reporting cycle. If the work is planned by the hour, then progress reporting should be cycled daily. Work planned on a daily basis is usually reported each week or two. Planning by the week is usually reported monthly. The reporting cycle must be longer than the time basis of the plan. On the other hand, planning by the hour but reporting only monthly is a waste of planning effort, or is being managed far too loosely. There will be more information on the criteria for review, updating, and reporting of project progress in Chap. 9.

Input needed. This is what drives the task. What is the input to the task which starts it? Each task is started by something, so identify it. This is a change in emphasis from traditional scheduling systems which demand work be completed by a given date. In project management, we place the emphasis on start dates. If a task starts on time and the work is accomplished within its estimated duration, the completion date will be satisfactory for the start of the next task. A task can need more than one input to start it.

From whom. This is linking information. It identifies which preceding tasks drive this task. This is the fundamental information which facilitates developing the logic diagram of the sequence of tasks in the flow of work and allows us to develop a schedule for each task.

The estimating step can take several weeks because several people must participate in the estimates. This is most efficiently accomplished by taking the WBS and estimating forms to the managers who will be responsible for the work. Those managers may want the people who would do the work to also do the estimating. All of this takes time and persistence.

Estimating errors. Estimating task durations and the amount of resources needed is subject to errors. These estimates have an element of judgment in them. If the task is a rerun of the same task in a previous project, the judgment element is low. If the task is a modification of previous experience, the need for judgment is higher. If the task is new or if the estimator has no previous experience with it, the risk of judgment error is quite high.

The source of estimating errors is our desires. If we are doing much estimating of tasks which requires a high judgment content from us, we must be aware of the source of our error. It comes from wishful thinking. If we want the task to be completed quickly, we have a tendency to estimate a shorter than realistic duration. Or, if we are short of people, we might estimate that two people can do the work which would normally take six. On the other hand, among the people who regularly do the work, there is a tendency to overestimate the duration and resource needs. This happens because their experience is usually that there is neither enough time to do the work nor enough people to get it done comfortably. Added to this tendency is the experience of the delay caused by unexpected problems, which also makes working people cautious.

These errors cannot be eliminated, but they can be recognized and compensated for. In a given situation, each of us is fairly consistent in our error. For example, an experienced senior project manager who always pushes for shorter schedules has a tendency to estimate durations low. The manager is aware of a certain consistency in the amount of underestimating he or she is guilty of. This person finds that by doubling the estimate and adding 50 percent, the adjusted estimate will be very close to actual performance. A 250 percent error would be unacceptably large, but by being aware of consistently erroneous behavior, the manager compensates and adjusts it to something realistic.

It pays for each of us to examine our estimating behavior. Accept it as the way we function. Then compensate for the error so that the estimate becomes more realistic. By compensating, we don't have to get involved in behavior change and personality problems.

TQM Task Estimates

In Chap. 5 we took the results of the TQA and applied the concept of the work breakdown structure. The WBS provided us with project phases, activities, and finally, tasks. The next step is to analyze the tasks and estimate their durations and the resources required. As a quick review, the tasks identified in the construction of the WBS were:

1.1.1 Write survey
1.1.2 Survey review/approval
1.1.3 Identify employees/departments
1.1.4 Schedule survey
1.1.5 Print survey
1.1.6 Conduct survey
1.2.1 Tabulate/review survey results
1.2.2 Approve communications media
1.3.1 Identify resources needed
1.3.2 Publish calendar
2.1.1 Review department operations
2.1.2 Review meeting
2.2.1 Supervisor subjects
2.2.2 Operator subjects
2.2.3 Determine class size/number
2.2.4 Schedule classes
2.2.5 Training materials
2.2.6 Conduct classes
2.3.1 Discuss with supervisors
2.3.2 Discuss with operators

2.3.3	Review charts
3.1.1	Identify consultants
3.1.2	Preliminary selection
3.2.1	Set appointments
3.2.2	Interviews
3.3.1	Review information
3.3.2	Group meeting

Each of these 27 tasks needs additional detail to fix responsibility and prepare the schedule. We will examine each task and assign the necessary items to accomplish this step. A brief description is also included for clarification of the logic used. The internal system found in Chap. 3, Fig. 3.4, is used for this example. Figure 6.2 represents the completed WBS worksheet.

Note: Total task durations are input on the worksheet, although resource requirements are listed individually. Except where noted, the longest resource duration is entered as the total task duration. In some cases, the resource durations may have to be added together to determine the total task duration. Ask yourself if the resources are required in series (add together) or parallel (longest resource duration = task duration).

Costs will be estimated for the resources using the following hourly rates:

Consultant	$125
President	$ 75
QA manager	$ 25
VP manufacturing	$ 40
Department mgrs.	$ 25
HR manager	$ 30
Supervisors	$ 20
Buyer	$ 25
Clerical	$ 10
Project owner	$ 25

We have tried to be as realistic as possible for the example, but we recommend that you think about it in terms of your own organization/departments/resources.

Project: TQM Implementation

Project Owner: C. P. Kaye

Date: January 1, 1993

Page 1 of 5

ITEM #	DESCRIPTION	RESPONSIBLE	RESOURCE	QTY	DURATION (DAYS)	INPUT NEEDED	FROM WHOM
1.1.1	Write Survey	Quality	Consultant	1	5	Hire Consult	2.1.1
			QA Manager	1			
			Human Resources Mgr	1			
1.1.2	Survey Review/Approval	Quality	Consultant	1	1	Write Survey	1.1.1
			QA Manager	1			
			Human Resources Mgr	1			
1.1.3	Identify Employees/Departments	Human Resources	Human Resources	2	3	Survey Approv	1.1.2
1.1.4	Schedule Survey	Human Resources	Department Managers	6	15	Identify Empl	1.1.3
1.1.5	Print Survey	Purchasing	PrintMasters	1	6	Survey Approv	1.1.2
1.1.6	Conduct Survey	Human Resources	Department Managers	6	10	Schedule Surv	1.1.4
						Print Survey	1.1.2
1.2.1	Tabulate/Review Survey Results	Quality	Consultant	1	3	Conduct Surv	1.1.6

Figure 6.2 TQM WBS worksheet.

Project: TQM Implementation

Project Owner: C. P. Kaye

Date: January 1, 1993

Page 2 of 5

ITEM #	DESCRIPTION	RESPONSIBLE	RESOURCE	QTY	DURATION	INPUT NEEDED	FROM WHOM
			QA Manager	1			
			Human Resources Mgr	1			
1.2.2	Approve Communications Media	Project Leader	President	1	1	Survey Results	1.2.1
1.3.1	Identify Resources Needed	Purchasing	Suppliers	?	5	Approve Media	1.2.2
1.3.2	Publish Calendar	Purchasing	PrintMasters	1	11	Approve Media	1.2.2
			Human Resources Mgr	1		ID Resources	1.3.1
2.1.1	Review Department Operations	Quality	Supervisors	3	5	Group Meeting	3.3.2
			Consultant	1			
2.1.2	Review Meeting	Quality	Supervisors	3	1	Operations Rev	2.1.1
			VP Manufacturing	1			
			Consultant	1			
2.2.1	Supervisor Subjects	Quality	Consultant	1	1	Review Meeting	2.1.2

Figure 6.2 *(Continued)*

Project: TQM Implentation

Project Owner: C. P. Kaye

Date: January 1, 1993

Page 3 of 5

ITEM #	DESCRIPTION	RESPONSIBLE	RESOURCE	QTY	DURATION	INPUT NEEDED	FROM WHOM
2.2.2	Operator Subjects	Quality	Consultant	1	1	Review Meeting	2.1.2
2.2.3	Determine Class Size/# People	Quality	Supervisors	3	1	Supervisor Sub	2.2.1
						Operator Subj	2.2.2
2.2.4	Schedule Classes	Quality	Supervisors	3	5	Class Size/#	2.2.3
2.2.5	Training Materials	Quality	Purchasing	1	9	Class Size/#	2.2.3
			Consultant	1			
2.2.6	Conduct Classes	Quality	Consultant	1	20	Schedule Class	2.2.4
						Training Mtls	2.2.5
2.3.1	Discuss With Supervisors	Manufacturing	QA Manager	1	2	Conduct Class	2.2.6
			Consultant	1			
			Manufacturing Mgr	1			
2.3.2	Discuss With Operators	Manufacturing	QA Manager	1	5	Conduct Class	2.2.6

Figure 6.2 *(Continued)*

Project: TQM Implementation

Project Owner: C. P. Kaye

Date: January 1, 1993

Page 4 of 5

ITEM #	DESCRIPTION	RESPONSIBLE	RESOURCE	QTY	DURATION	INPUT NEEDED	FROM WHOM
			Supervisors	3			
			Consultant	1			
			Manufacturing Mgr	1			
2.3.3	Review Charts	Quality	QA Manager	1	10	Supervisor Dis	2.3.1
			Consultant	1			
			Supervisors	3			
			Manufacturing Mgr	1			
3.1.1	Identify Consultants	Quality	QA Manager	1	5	TQM Approval	0.0.0
3.1.2	Telephone Screening	Quality	HR Manager	1	5	ID Consultants	3.1.1
3.1.3	Preliminary Selection	Quality	HR Manager	1	3	Telephone Scre	3.1.2
			QA Manager	1			
3.2.1	Set Appointments	Human Resources	HR Manager	1	5	Prelim Select	3.1.3

Figure 6.2 *(Continued)*

Project: TQM Implementation

Project Owner: C. P. Kaye

Date: January 1, 1993

ITEM #	DESCRIPTION	RESPONSIBLE	RESOURCE	QTY	DURATION	INPUT NEEDED	FROM WHOM
3.2.2	Perform Interviews	Quality	President	1	10	Set Appts	3.2.1
			QA Manager	1			
			Project Owner	1			
			HR Manager	1			
3.3.1	Review Information	Quality	President	1	1	Perform Interv	3.2.2
			QA Manager	1			
			Project Owner	1			
			Hr Manager	1			
3.3.2	Group Meeting	Quality	President	1	1	Review Info	3.3.1
			HR Manager	1			
			QA Manager	1			
			Project Owner	1			

Figure 6.2 *(Continued)*

1.1.1 Write survey

Responsible: Quality

Resource/quantity/duration (days)

Consultant/1/2 $500
QA manager/1/5 $300
HR manager/1/2 $100

Input needed/from: Group meeting/3.2.2

The quality department has had the most experience in writing surveys and, therefore, will be the responsible party. The quality manager will be required to devote the most time because of that experience, while the consultant and human resources manager will provide input. This task cannot begin until the consultant is hired.

1.1.2 Survey review/approval

Responsible: Quality

Resource/quantity/duration (days)

Consultant/1/1 $125
QA manager/1/1 $ 25
HR manager/1/1 $ 30

Input needed/from: Write survey/1.1.1

This is the subsequent task to item 1.1.1, formalizing the survey. Responsibilities and resources are identical, but the duration is considerably less due to the nature of the task.

1.1.3 Identify employees/departments

Responsible: Human resources

Resource/quantity/duration (days)

Human resources/2/3 $120

Input needed/from: Project approval/0.0.0

The HR department should maintain all of the current employee records, so this is the most likely area for responsibility and action on the task. We will assign two people for three days to gather the information. This task cannot start until the project is approved.

1.1.4 Schedule survey

Responsible: Human resources

Resource/quantity/duration (days)
Department managers/6/15 $1500

Input needed/from: Identify employees/1.1.3

The HR department should work with all of the department managers (R&D, quality, purchasing, etc.) to establish the time frame for the survey. The department managers will administer the survey within a 15-day window. Because production managers tend to have little time for other items, we have allowed 15 days to finalize the schedule. This task cannot begin until the breakdown of employees by department is available.

1.1.5 Print survey

Responsible: Purchasing

Resource/quantity/duration (days)
PrintMasters/1/6 $1000

Input needed/from: Survey approval/1.1.1

Since the purchasing department's function is the procurement of goods and services, the responsibility for the task will be here. PrintMasters is the printer, who requires two days for processing, and four days for scheduling/setup/printing. This task cannot start until the survey is approved.

1.1.6 Conduct survey

Responsible: Human resources

Resource/quantity/duration (days)
Department managers/6/10 $600

Input needed/from
Schedule survey/1.1.4
Print survey/1.1.5

Although the department managers are going to administer the survey, it is still the responsibility of the HR department to ensure task completion. The 10 days scheduled for this task includes any makeup sessions which might be required. In this case, the survey cannot start until *two* other tasks are completed: the survey schedule and printing.

1.2.1 Tabulate/review survey results

Responsible: Quality

Resource/quantity/duration (days)

Consultant/1/1	$500
QA manager/1/3	$300
HR manager/1/1	$120

Input needed/from: Conduct survey/1.1.6

The QA manager will tabulate the results with assistance from the HR manager and the consultant. The total duration for the task is three days. This task cannot start until the survey is completed.

1.2.2 Approve communications media

Responsible: Project leader

Resource/quantity/duration (days)

President/1/1 $150

Input needed/from: Survey results/1.2.1

The project leader will be responsible for getting the approval of the company president. In this case, the president wanted to be actively involved because of the companywide impact. One day is set aside, as the president is readily available (especially on quality issues). This task cannot start until the survey results are available.

1.3.1 Identify resources needed

Responsible: Purchasing

Resource/quantity/duration (days)

Suppliers/?/5 $250 (buyer)

Input needed/from: Approve communications media/1.2.2

Purchasing is in the best position to identify the suppliers required for the communications media selected. The number of suppliers will not be known until item 1.2.2 is complete. For instance, the resources for an in-plant video communications system will be quite different from a one-page typewritten newsletter.

1.3.2 Publish calendar

Responsible: Purchasing

Resource/quantity/duration (days)

PrintMasters/1/6 $1000
HR manager/1/5 $ 150

Input needed/from

Approve communications media/1.2.2

Identify resources/1.3.1

Purchasing will be responsible for this task. The HR manager will develop the calendar and, again, PrintMasters will perform the printing. This is a case where the resources are needed in series, resulting in the addition of the duration estimates. The calendar cannot be printed until the events on it are determined by the HR manager. This task cannot be started until the two tasks, resource identification (item 1.3.1) and media approval (item 1.2.2), are completed.

2.1.1 Review department operations

Responsible: Quality

Resource/quantity/duration (days)

Supervisors/3/5 $ 720
Consultant/1/5 $5000

Input needed/from: Group meeting/3.2.2

This task requires the involvement of three department supervisors and the consultant. A review of the operations and discussions with the supervisors over a five-day period will take place. This task cannot start until the consultant is hired. Because of the nature of the task, statistical methods, the quality department is responsible.

Chapter

7

Step 5: Calculate the Schedule with a Logic Network

Building the Model

In step 4, we collected the raw data for our scheduling model. In step 5, we build and optimize that model. It usually amazes newcomers to project management that we don't have a schedule until we are five steps into the process. This occurs because there is a lot of work to be done before we can develop accurate schedules. Project management schedules are based on the best estimates we can get. Most other scheduling systems base their dates on wishful thinking, which is characterized by the phrase, "I want . . ." Such scheduling systems are especially vulnerable to the estimating errors covered in Chap. 6.

At the completion of step 4, we had estimates of all the tasks. The sum of these durations will provide us with a calculation of the time input into a project, but it will not give us a schedule for the sequence of work. For example, if the task durations were plotted in a chart as shown in Fig. 7.1, they would all start at the first day because we have no way of showing the precedence of one task over another. What we are trying to do is get to the method of scheduling which explains how task durations are stair-stepped and overlapped as shown in the Gantt chart of Fig. 7.2.

Project management software

Project management software has been the subject of many books. There is more opinion on the subject than there are peo-

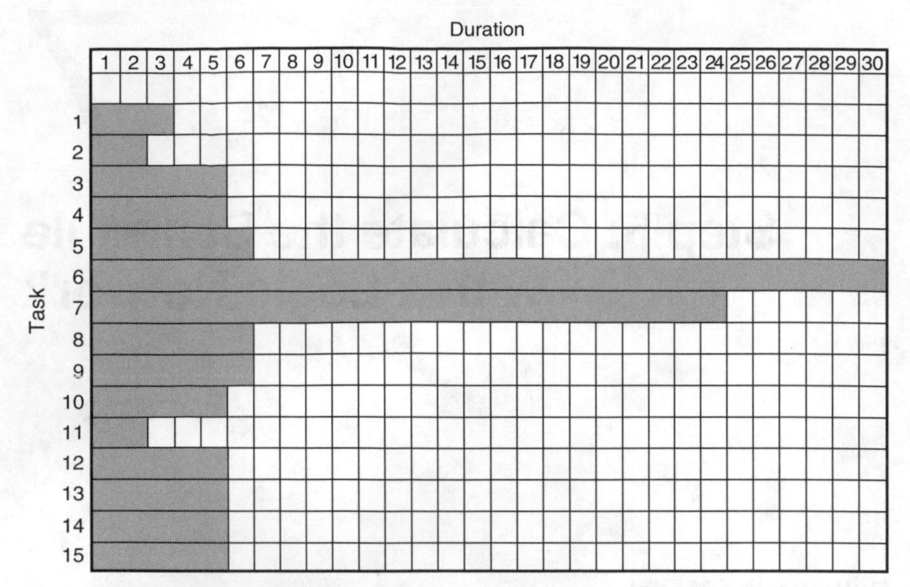

Figure 7.1 Project management duration bar chart.

Time

	Time	5	10	15	20	25	30	35	40	45	50	55	60	65	70	75	80	85	90	95	100
Close safety gates	3																				
Clear	2																				
Pull loader	5																				
Pull unloader	5																				
Turn trans	6																				
Clean	30																				
Move out and change rails	24																				
Unbolt press	6																				
Out-in	6																				
Tie up new	5																				
Install trans	2																				
Bottom and connect hoses	5																				
Install unloader	5																				
Adjust loader	5																				
Final adjustment	5																				

Figure 7.2 Project management Gantt chart.

ple working in project management. The fundamental facts are that there are more than 200 project management software packages available on the market, and there is no perfectly satisfying one in the bunch. So when you hear someone proselytizing on the superior virtues of one brand over the others, you are probably hearing the words of a novice or a salesperson. Most professionals who work with project management software regularly, and who work with several brands, are frustrated by the deficiencies of each.

The reason for these persistent deficiencies is not the blame of the software developers and publishers. Project management software is complex and must have a broad range of human interfaces. To further complicate matters, the underlying dynamic philosophy of project management is not understood by most managers. It is a flow concept in which work is seen as moving from one operation to another. This idea of flow requires a change in viewpoint for those managers who see a snapshot of work passing the point where they are sitting. This snapshot viewpoint is not a deficiency of those managers, because we are taught to visualize that way.

The complexity of project management software is what has spawned so many brands. Each has its own solution to the human interface problems. As you examine different software packages and listen to salespeople, remember that all of the packages perform the same fundamental function—calculate task- and time-based scheduling. Many of the operations are similar. Some standardization has been introduced by the Project Management Institute. The differences are apparent, mostly in the human interfaces—the displays that allow us to control operation of the software and how results are displayed. The human interfaces are the source of what is commonly referred to as user friendliness.

If you are selecting project management software, the important criteria are how skilled you expect to become in operating it and how much flexibility and control you are going to need. An engineer who is going to run it once a month needs more user friendliness than a professional who will run it three times a day for different customer requirements. The professional will emphasize flexibility and speed.

A plotter is a necessary piece of equipment for viewing network logic and preparing Gantt chart reports. Many of the more useful outputs from calculating a critical path method (CPM) network are graphical.

Logic conventions

In the 1950s and 1960s, we did logic network scheduling manually without computers. Today, the universal availability of computers and project management software for a few hundred dollars have made these the overwhelming choice. They have also made complex logic relationships such as start-to-start and finish-to-finish more common, and manual calculations with these is very difficult.

The logic network diagram is a systematic and disciplined approach to organizing and displaying all the tasks of a project, their sequence, and relationships to each other. It presents the logical sequence of tasks that must occur to complete the project within the allocated time and budget.

The logic network and its operation and use make up the critical path method (CPM). Therefore, CPM refers to the computer-aided scheduling system which generates early and late start and finish dates for each task, based on the links between tasks. Output of the results of the calculation is usually in tabular data tables or graphical bar charts, histograms, and logic network diagrams in a variety of formats.

When entering the data from step 4, the software will expect us to enter the logic between tasks. Today, this is represented by the precedence diagramming method (PDM), which uses boxes to represent tasks (see Fig. 7.3) and connecting lines to repre-

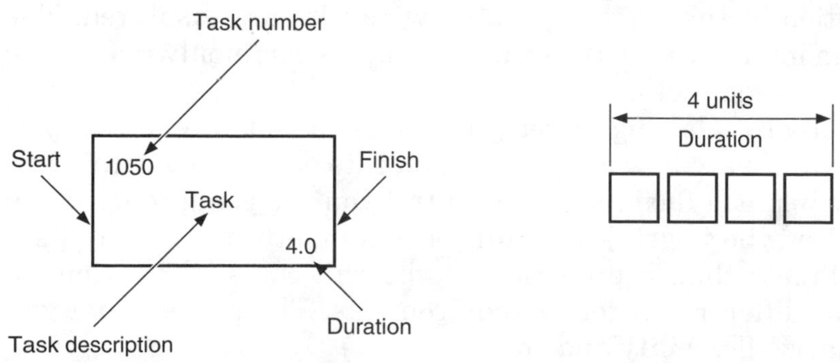

Figure 7.3 Basic logic elements.

sent logic (see Fig. 7.4). For each task in the WBS there is almost a 1:1 translation to the boxes in the logic network. The task description, duration, resources, and predecessor task links are entered for each box. Thus, the network grows from box to box. The linking of tasks is what makes our model a simulation of the flow of work for our project. When all of the tasks are entered and linked, the model can be run to calculate the schedule for the entire project as well as for each task. The calculated start and finish date for each task will appear in the task box.

In PDM, three types of logic links can be used to represent the relationships between tasks, as shown in Fig. 7.5. These logic links are sometimes referred to as constraints because they impose a constraint on the start or finish of the task to which they are connected.

1. *Finish-to-Start (FS)* links are the most common, and are the default link in most project management software. The start of task B is constrained by the finish of task A; that is, task A must finish before task B can start. This represents the project relationship where task B must have the output of task A before it can start.
2. *Start-to-Start (SS)* links constrain the start of task B to some time after the start of task A. There is usually a delay from the start of task A until task B can start. It is usually stated in percent of completion of task A, or in hours, days, or weeks

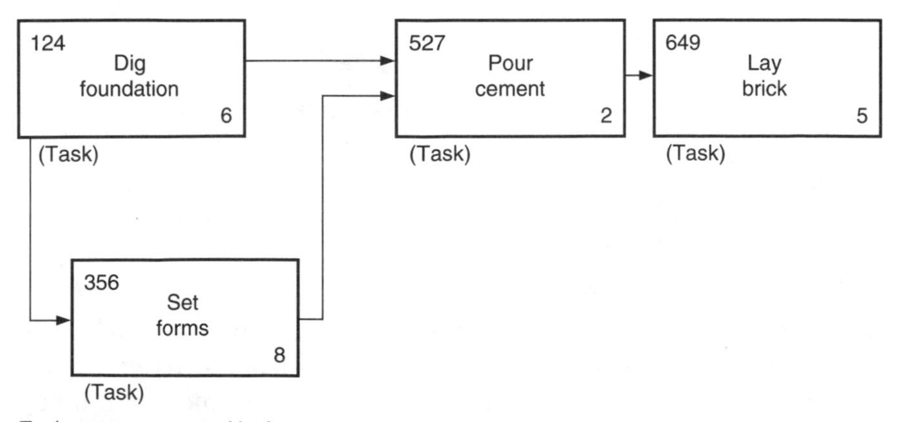

Figure 7.4 Precedence diagram method (PDM).

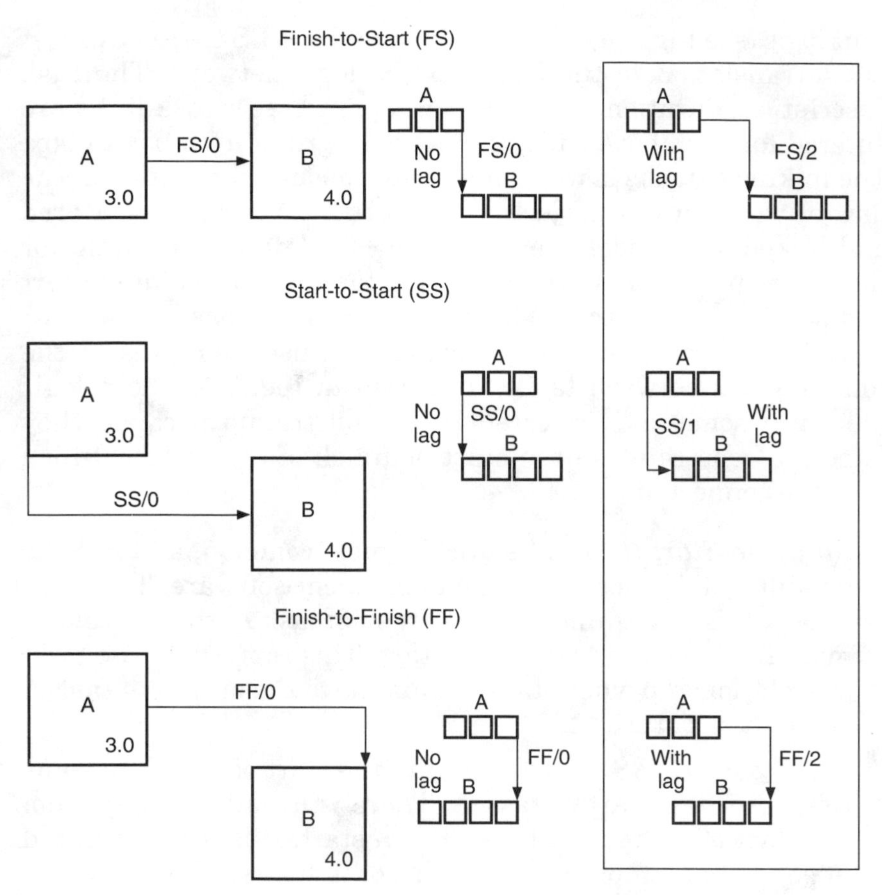

Figure 7.5 Types of logic relationships.

after the start of task A. This represents the project relationship where task B gets some kind of input from task A while task A is in progress. The finish of task A is not important to task B. If task A were testing and task B were final design, final design (task B) would not have to wait for the completion of testing (task A) before it started.

3. *Finish-to-Finish (FF)* links constrain the finish of task B to the finish of task A. Regardless of when either started, task B cannot finish until task A has finished. There may be a delay between the finishes. In the project, this represents the relationship where task A must finish before task B can finish. For example, a product certification application cannot be

completed before testing is completed because the test results must accompany the application.

Early dates versus late dates. In a CPM logic network, initial calculations are made from beginning to end (graphically, from left to right). CPM logic networking rules do not permit feedback loops as in system flow charts. CPM logic networks are process networks with an implied time base. Consequently, the network can be calculated in sequence regardless of the number of paths through it. A simplified network is shown in Fig. 7.6.

Calculations in a simplified logic network like that shown in Fig. 7.6 are performed in two parts. Calculations are first made from left to right, in the direction of the flow of work, as shown in Fig. 7.7. This is referred to as the *forward pass.* The dates calculated for each task are called the *early dates* because each date is the earliest its task can be started and finished. Early finish dates are calculated by adding the task's duration to its early start date. Its early start date is the finish date of its predecessor task. The early dates are shown above the task boxes in Fig. 7.7.

The initial task starts at 0 days, or the project start date. The early finish date of task 1 is calculated by adding its duration of 1 day to its start date of 0. This gives task 1 an early finish date of 1. Task 2 then starts on the same day as its predecessor, task 1, finished—day 1. The early finish date for task 2 is the sum of its duration, 4 days, and its start date. This gives task 2 an early finish date of 5.

When there is more than one predecessor task, the task cannot start until the latest predecessor finishes. For example, task 6 cannot start until task 3 finishes on day 11, even though preceding task 4 finished on day 9.

The finish date of the last task, task 9 in Fig. 7.7, is the projected finish date of the project. In a complex project, there may be several last tasks, one at the end of each path. In that situation, the project finish date is the latest finish date of these last tasks.

Once the forward pass calculations are made, the logic network can then be calculated for *late dates.* These are the latest start and finish dates each task can have without jeopardizing the project completion date. Late start and finish dates are calculated by making a backward pass through the logic network as shown in Fig. 7.8. The late dates are displayed below the task boxes.

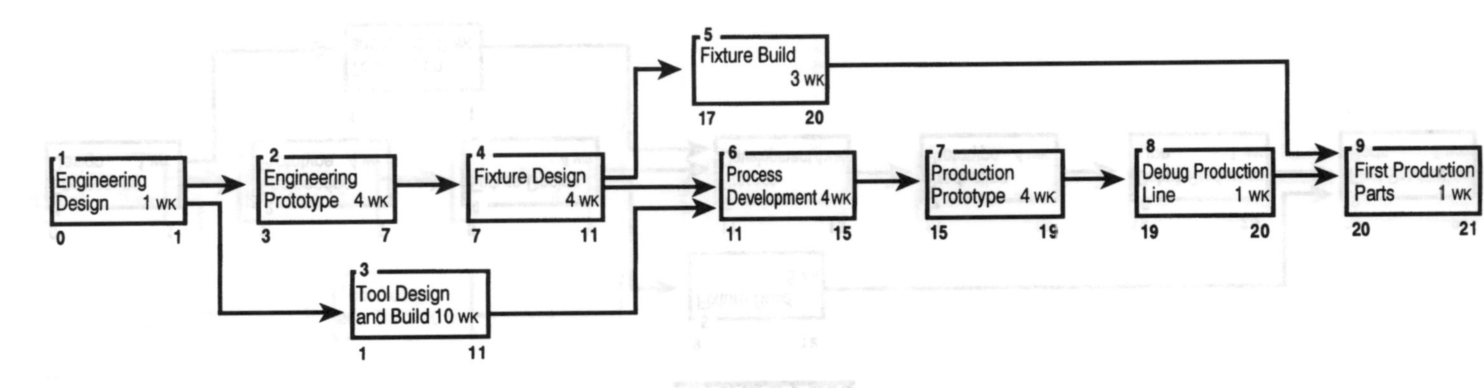

Figure 7.8 Backward pass.

Backward pass calculations are made from right to left in the logic network. The late start date of each task is calculated by subtracting the task's duration from the late finish date. The late finish date for its preceding task is the same as this task's late start date. For example, in Fig. 7.8, the calculation of late dates starts with day 21, the date calculated for project finish in the forward pass. The late start date for task 9 is day 21 minus its 1 day duration, day 20. Task 8 then has a late finish date of day 20. Its late start date is day 20 minus its duration of 1, day 19.

Where there are multiple paths into a task in the backward pass, as for task 4, the earliest late start date, day 11, is the latest task 4 can finish.

Early and late dates are integral to the calculation of two important types of scheduling information: float and critical path. Float is calculated as the difference between the early and the late dates for a task. Either the start dates or the finish dates can be used. Where the difference (float) is 0, the task is a critical task on the critical path. If the difference is a positive number, there is float between the early and late dates, and the task is noncritical.

Float can be calculated as negative when the task durations are too long to fit between constraining dates, such as between sign-off for a design and the planned introduction date for the product.

This condition is referred to as supercritical because it tells us that we cannot complete the project on time with this plan. As a word of caution, some project management software produces unreliable calculation results in this situation.

If the logic network is built with start-to-start and finish-to-finish logic, the preceding type of manual calculation becomes quite difficult. The computer software, however, seems to handle this calculation easily.

Critical path. The critical path of a logic network is that string of tasks which comprise the longest path through the model. The term *critical* in project management has nothing to do with the importance of the task. It refers to the task being schedule-critical, since the project cannot be completed any sooner than its critical path. To quote my good friend and respected consultant Jim Rouhan, "The longest is the shortest!"* That is, the longest path through the logic network is the shortest time in which the project can be completed.

Noncritical tasks and float. Critical tasks are the tasks along the critical path. They have no float or slack time. That is, they must be completed by their estimated completion dates or the completion of the project will be delayed. Noncritical tasks, on the other hand, have some scheduling leeway, called *float.* Float is the window of time in which the task can be performed. This window is longer than the duration of the task, so the scheduling for the task can float back and forth within its float window without delaying the next task. This window, associated with each noncritical task, is referred to as *free float.* As long as the performance of the task stays within this window, it will not delay the tasks following it.

Another term, *total float,* refers to the total amount of float for the project which exists at the scheduled time of each task. We have to be careful about which type of float is being displayed. No one gets upset about a task using its free float. But downstream operations get very upset about a task using all the total float. For example, manufacturing people get very upset about engineers using all of a project's total float. The hogging of float can make all subsequent tasks go supercritical; that is, their floats become negative, and the critical tasks are the most negative. In the project, this situation means that if these tasks take their planned durations, the project completion will be late.

Optimizing the schedule

The initial run of a schedule invariably needs optimizing because it shows completion much later than desired. The challenge is to take as much time out of the schedule as practical. There are several trade-offs which can be made to shorten the planned time, and there is a sequence of actions to do this:

1. *Identify the critical path.* It is composed of the tasks which threaten to make the project finish late (see Fig. 7.9). Shortening noncritical tasks and paths will not be as effective. As

---

* Rouhan, J., Project Management Planning and Tracking Workbook, Management Technologies, Inc., 1989, sec. V, p. 19.

OPTIMIZING THE SCHEDULE:

Step 1: Identify the Critical Path

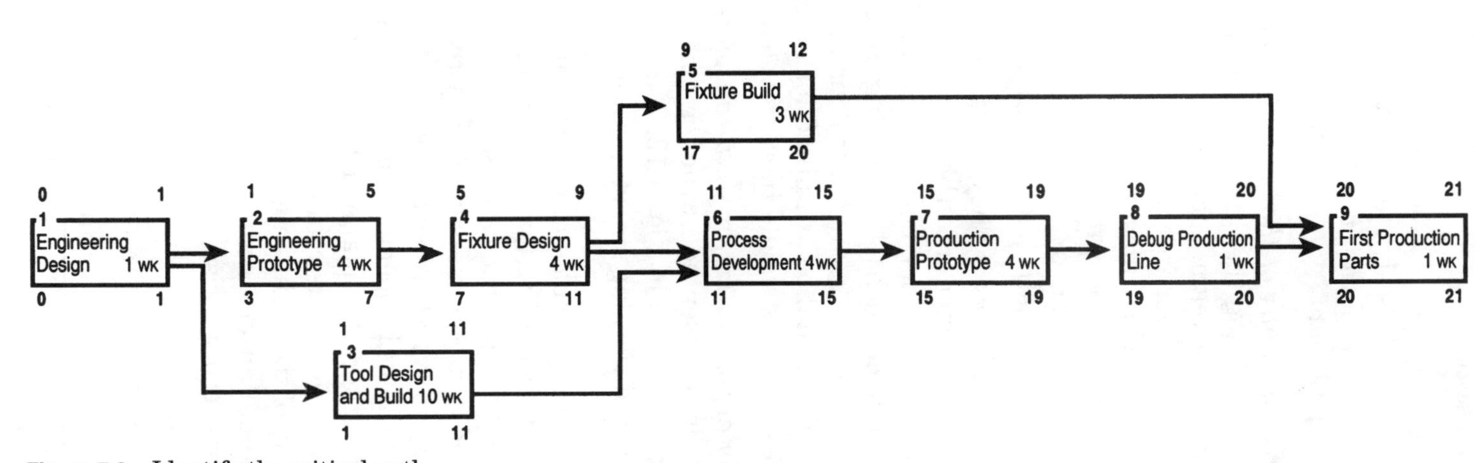

Figure 7.9 Identify the critical path.

the critical path is shortened, other paths may become critical. The critical path tends to move around during optimization. Keep track of it after each operation.

2. *Compress as many of the critical tasks as practical.* This is done by reviewing the duration estimates with the responsible manager or team. Increasing work-force allocations may shorten some. Overtime may shorten others. Changes in process and technology may shorten some. Take as much time as possible out of the critical path, but do it in increments, because the critical path may shift to another set of tasks. This review is illustrated in Fig. 7.10.
3. *Identify bottlenecks.* They always slow work, but they are sometimes designed into the system as reviews and sign-offs (see Fig. 7.11). Other bottlenecks can be caused by insufficient resources, such as machining. Handoffs are a dangerous form of bottleneck because they can stop work if fumbled. All bottlenecks should be examined for necessity and delay time. Unnecessary ones should be removed.
4. *Look for opportunities for simultaneous effort.* Simultaneous activity is represented in the logic network as parallel paths of tasks, as shown in Fig. 7.12. This is project management's definition of working smarter, not harder.

A lot of faith was put into this technique in the auto industry, but it did not always live up to its promise. The failures seemed to come from two general causes: failure to communicate more actively and failure to manage more actively. Simultaneous activity can shorten overall project time, but it does not come free. Having several things happening simultaneously which are supposed to come together at a planned time creates a big management load. A lot of cowhands could ride one horse well, but few could drive a team of six. Additional management effort is the price to be paid for simultaneous activity. If you can't pay the price, don't make the move; otherwise, you'll just get into bigger trouble faster.

Resource leveling

Resource leveling is a feature included in many project management software packages. Its purpose is to push peak work-force overloads into work-force valleys, thus smoothing the work-force

TASK NO.	DESCRIPTION	ORIGINAL DURATION	REVISED DURATION
1	Engineering Design	1 WK	2 WK
3	Tool Design & Build	10 WK	8 WK
6	Process Development	4 WK	3 WK
7	Production Prototype	4 WK	4 WK
8	Debug Production Line	1 WK	1 WK
9	First Production Parts	1 WK	1 WK

Figure 7.10 Identify critical tasks which can be compressed.

OPTIMIZING THE SCHEDULE:

STEP 3: Identify Bottlenecks, Examine Their Purpose, Eliminate or Accelerate Them if Practical

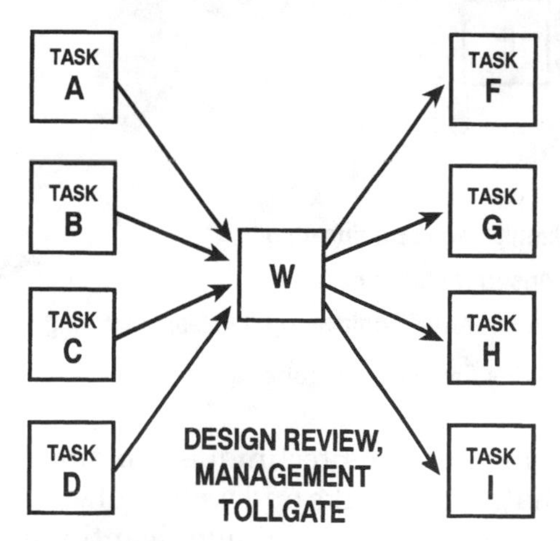

Figure 7.11 Identify bottlenecks.

requirements. Resource demand is usually displayed on a histogram for each resource identified in the project. Figure 7.13 shows an example of a resource histogram.

Resource leveling is accomplished by moving noncritical tasks within their float windows to spread out the demand. Within the

OPTIMIZING THE SCHEDULE:

STEP 4: Identify Opportunities for Simultaneous Tasks

• **Accelerate:**

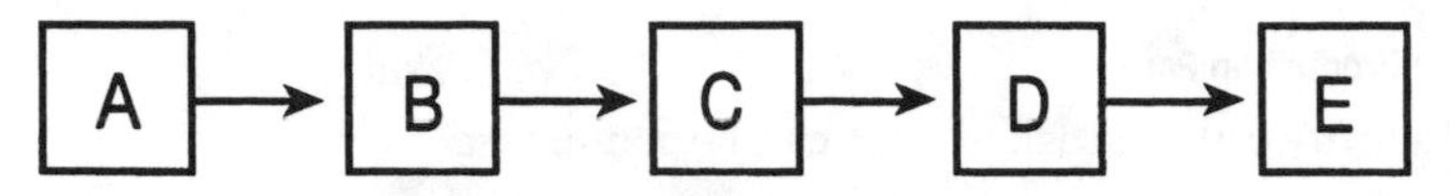

• **By Simultaneous Work**

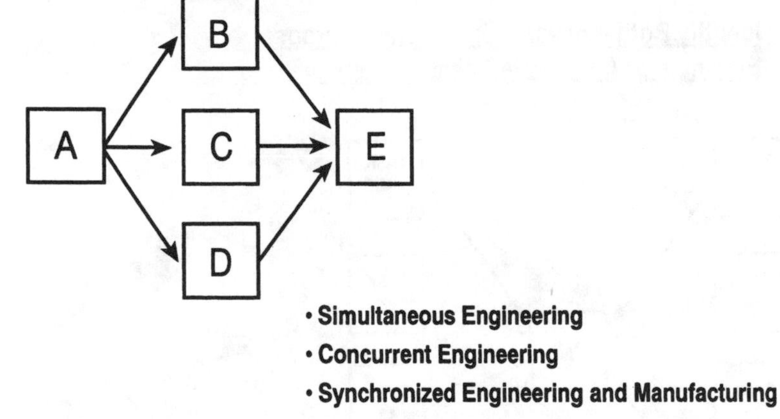

Figure 7.12 Identify opportunities for simultaneous tasks.

limitations of the float windows, this technique can smooth the resource demand somewhat. If the peaks and valleys of work-force demand are large, the variations may not be significantly changed by adjusting noncritical tasks. The alternative, then, is to adjust the timing of all tasks, including critical tasks. This approach pushes out the project completion date. When doing resource leveling with software, always check that the project completion date was not extended. In most industrial development projects, extending the project completion date is not acceptable; therefore, another solution to peak work loads must be found.

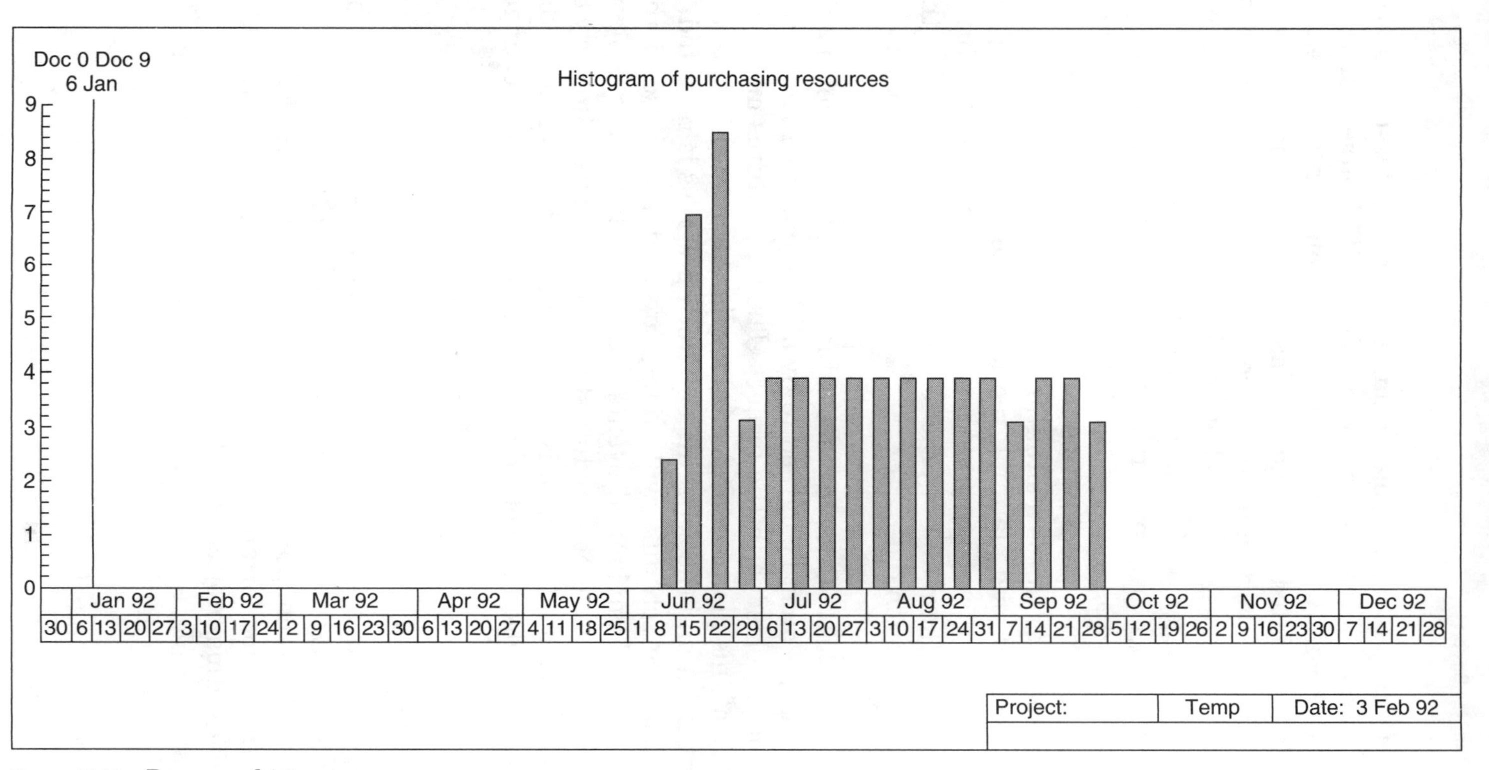

Figure 7.13 Resource histogram.

Line managers and project management professionals frequently run resource histograms on their initial project plans to display the peak work-load points. Most software will identify the tasks associated with the overloads. With this information, the manager usually runs resource leveling on the noncritical tasks to quickly see what relief can be gotten from float time. Then the manager begins to examine the tasks which are contributing to each overload, making procedural changes in the project to relieve as much of the overload as possible. The manager also examines possibilities for temporarily increasing overloaded resources. This is a situation where creative management is really needed, and planning reduces trauma. By juggling the flow of work and temporary help, the manager can usually bring overloads into the realm of practicality without jeopardizing the project's completion date.

Baseline schedule. Once a reasonable plan has been developed by the team, it must be agreed to by all line managers and project people involved with the project. Their signatures make the schedule of this plan the project's baseline schedule. This is the starting point and the schedule on which progress is reported.

The baseline schedule will change over time because of the unforeseen and unplanned events during the life of the project. Changes to the baseline schedule should be controlled similarly to changes to design drawings. Because progress is reported to the baseline schedule, all involved parties must agree to each change.

TQM Scheduling

Our next step in the TQM implementation process is to take the task duration estimates and begin to develop the schedule. The first part of this chapter demonstrated the mechanics of the critical path method for any project. It can be applied to the TQM implementation project as well.

In our example, we will use the following sequence of events to create our logic network:

1. Convert/arrange tasks to PDM boxes.
2. Show interrelationships between tasks.

3. Evaluate linkages between tasks.
4. Use the forward pass to calculate early starts and finishes.
5. Use the backward pass to calculate late starts and finishes.
6. Determine the critical path.
7. Evaluate the float.
8. Look for opportunities of optimization.

When we have completed these eight scheduling steps, the road map will have been established for the project in terms of time. This will be used to determine performance for the duration of project.

The use of project management software and associated problems were discussed at length in the first part of this chapter. The continuing example of the TQM implementation is illustrated first manually, then using a software program. This will give the reader a better appreciation of both manual preparation and the necessity of software. Remember, the example requires immediate attention to only two areas, as determined by the TQA. Imagine the work required to manually prepare the schedule for 10 or 15 areas in a larger company!

1. Convert/arrange tasks to PDM boxes. The first step of our scheduling process is to convert all of the tasks, identified in the WBS worksheet, to PDM boxes. This means identifying the task in two ways: first, by placing the task number in the upper-left corner of the box, then by describing the task in the center of the box. In the lower-right corner of the box, we will put the estimated task duration. As this is being done, we can also arrange the boxes in an approximate time sequence from left to right. For example, we know that task 1.1.6 ("Conduct survey") will not take place before task 1.1.1 ("Write survey"). It must be somewhere downstream of writing the survey. Try to identify the tasks which must occur first, and work your way to the right. Figure 7.14 shows all of the logic elements arranged in approximate time sequence, from left to right. A linear arrangement of the 22 tasks would be the easiest, but least effective, method. Wherever possible, arrange several tasks to be performed simultaneously. However, be careful not to schedule one resource for multiple tasks at the same time.

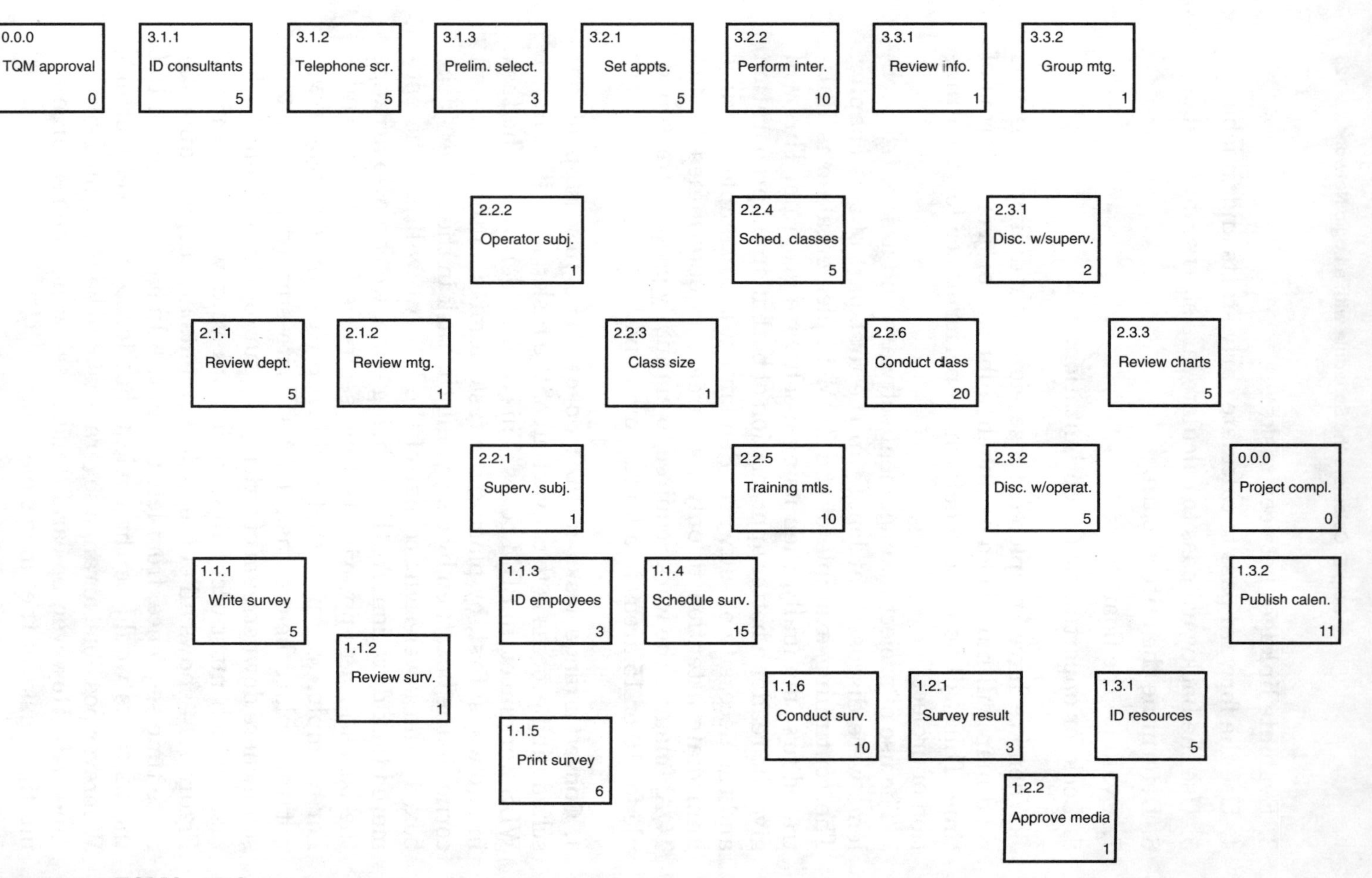

Figure 7.14 TQM logic elements.

2. Show interrelationships between tasks. Now that the tasks have been arranged in a logical order, we need to develop the relationships between them. The addition of lines with arrows indicating the flow of work from task to task is the beginning of our precedence diagram. Figure 7.15 illustrates the connections between the tasks. All of the tasks in our TQM example are preceded or followed by one or two tasks. TQM implementations with a broader scope would most likely have tasks preceded or followed by more than two tasks. For instance, following task 2.2.6 ("Conduct classes") we might also want to review the classes with top management. In that case, three tasks, 2.3.1, 2.3.2, and "Discuss with top management" would follow 2.2.6 and precede 2.3.3.

3. Evaluate linkages between tasks. All of the tasks in the TQM example require the completion of the previous task to begin the current task. This is called a *finish-to-start linkage.* (Refer to the first part of this chapter if necessary.) The two other types of linkages are *finish-to-finish* and *start-to-start.* Even though we used the finish-to-start linkages, there may be places in the diagram where one of the other two might be appropriate. If we examine the two tasks, 1.1.3 ("ID employees") and 1.1.4 ("Schedule survey"), another linkage may be possible. Scheduling of the survey could begin as the various departments identified and reported their employees. The smaller functions would complete the assignment first and may be scheduled first. Therefore, a start-to-start linkage may also be as good as or better than the finish-to-start linkage.

4. Use the forward pass to calculate early starts and finishes. As in the first part of this chapter, the early starts and finishes of each task must be calculated. Again, we work from left to right, adding the duration of the following task to the end point of the preceding task. For example, task 1.1.6 has an early start of 54 days and a duration of 10 days. Therefore, the early finish of this task is 64 days. The following task, 1.2.1, then has an early start of 64 days, which is the early finish of task 1.1.6. Task 1.2.1 has a duration of 3 days and will have an early finish of 67 days. In a situation where a task is preceded by two tasks and the linkage is finish-to-start, the task with the longest duration is used

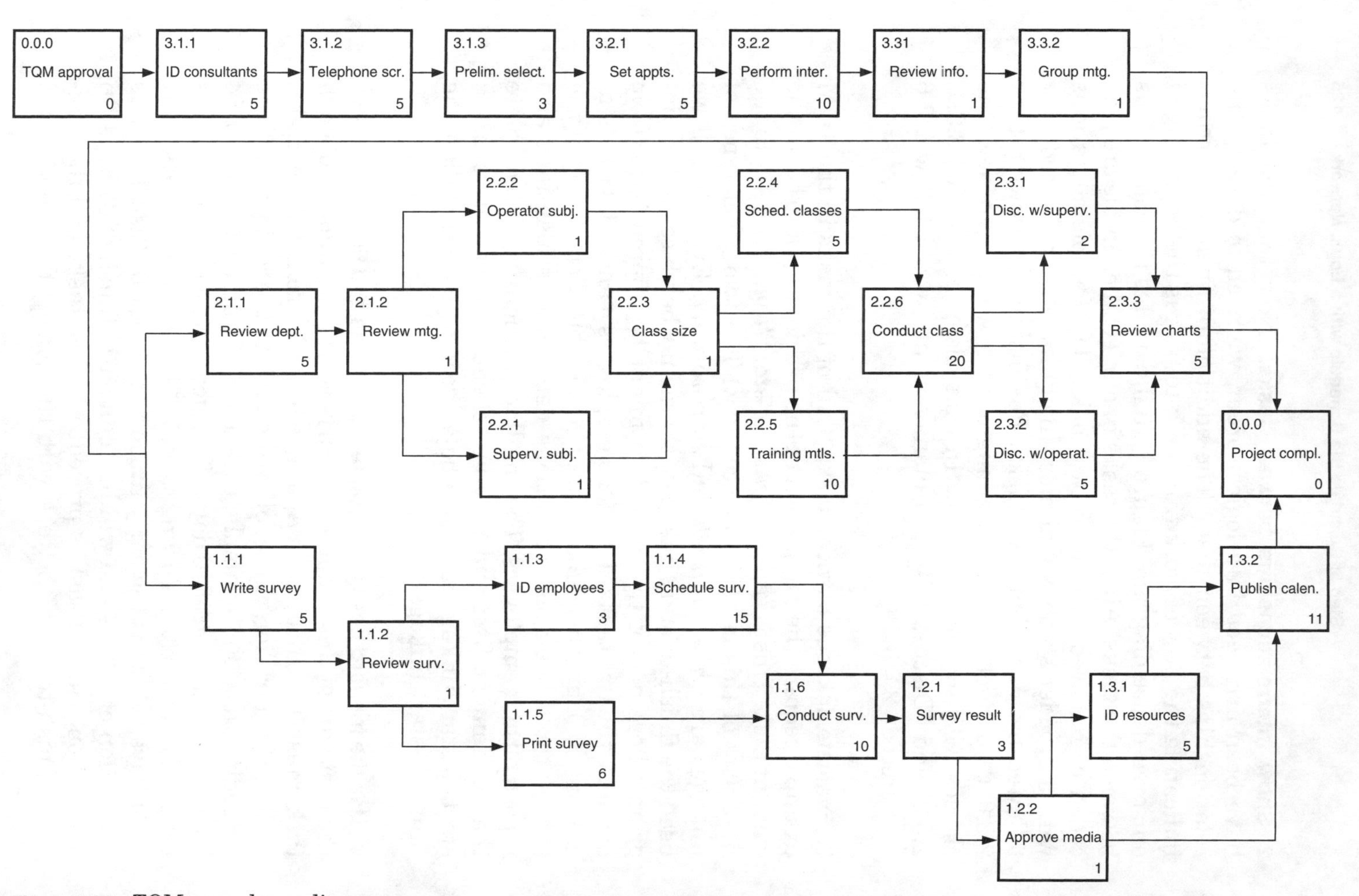

Figure 7.15 TQM precedence diagram.

in the calculation. Tasks 2.2.4 and 2.2.5 both have the same early start date; however, the duration of the tasks is different. Task 2.2.5 is the longer task and is used to calculate the early start of the following task, 2.2.6. Figure 7.16 shows all the tasks with their early starts and finishes.

5. Use the backward pass to calculate late starts and finishes. The late starts and finishes of the tasks are calculated in much the same way as the early starts and finishes. In the case of late starts and finishes, we will begin from the right of the diagram and work to the left. The total duration for the project is 84 days, which is used as the starting point for the backward pass calculations. Figure 7.17 shows the late starts and finishes for each task. In situations where two tasks flow back into one, as in tasks 1.1.3 and 1.1.5, the smallest late start is used for the calculation of task 1.1.2. When the late starts and finishes of all the tasks are completed, we can begin to evaluate the project schedule in its entirety.

6. Determine the critical path. After having calculated both the early and late starts and finishes, the next step is going to seem quite simple. Remember the expression, "The longest is the shortest," from the first part of this chapter? The longest path through the schedule is the shortest amount of time the project can be completed, which is the critical path. Figure 7.18 shows all the tasks, the early and late starts and finishes, and the critical path. The bold lines and arrows through the tasks represent the critical path. Comparing these tasks to the others not on the critical path, you will immediately notice that the early starts and finishes equal the late starts and finishes. Very simply, this means that a delay in any task on the critical path will delay the project. For instance, it might take longer than 30 days to find and hire a consultant for the project. If it takes 35 days instead of the 30 days originally planned, the total duration of the project is automatically extended to 89 days.

7. Evaluate the float. In evaluating the float within a project, we must observe both the total float and the free float. Figure 7.18 shows us several examples of both types of float. Studying the

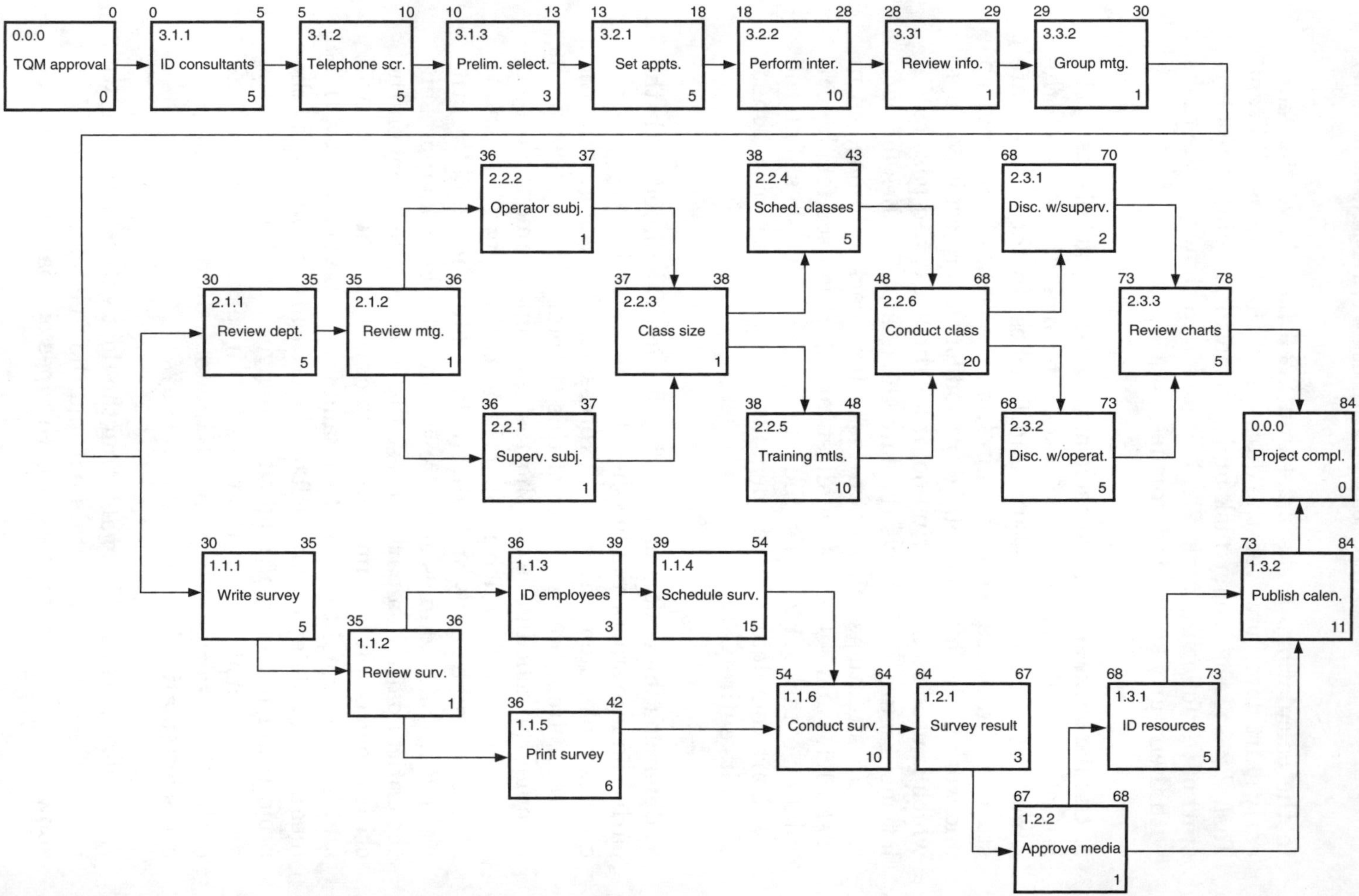

Figure 7.16 TQM forward pass.

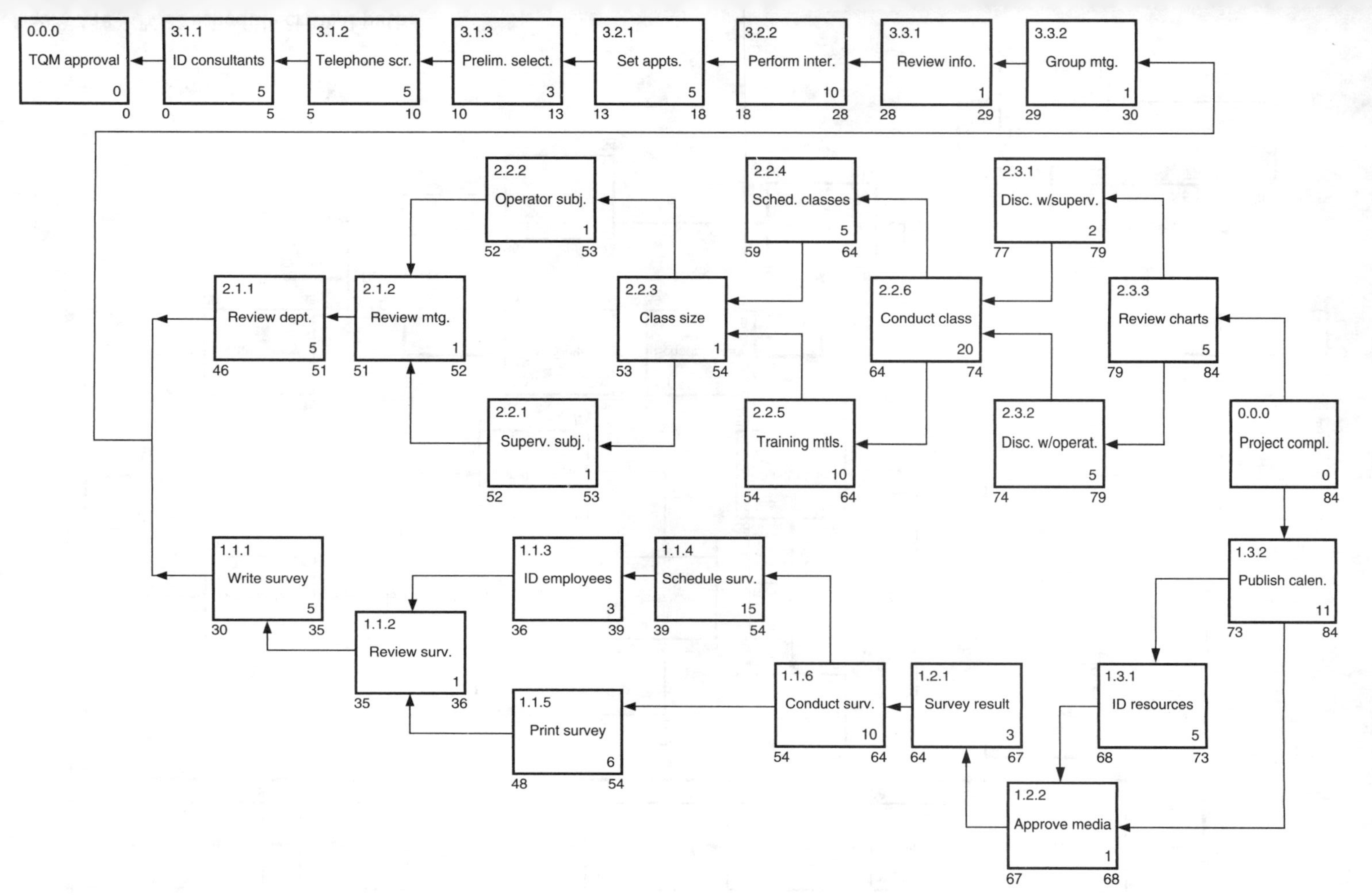

Figure 7.17 TQM backward pass.

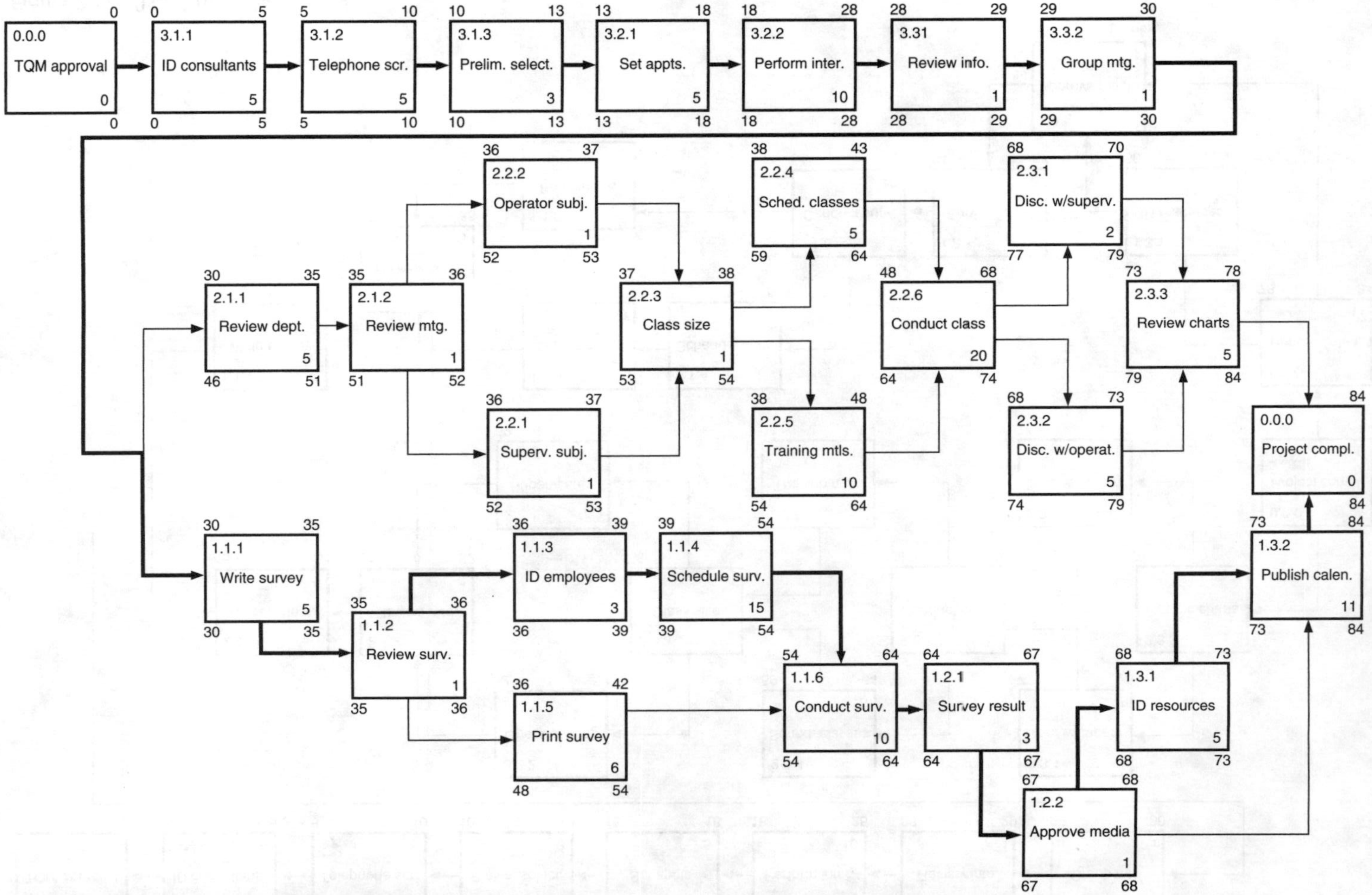

Figure 7.18 TQM schedule critical path.

float of a project allows us to make decisions (or at least be prepared) with regard to the schedule.

Tasks 2.2.4 and 2.2.6. To determine the total float of task 2.2.4, we need to subtract the early start from the late start, or the early finish from the late finish. This gives us 59 – 38 = 21 or 64 – 43 = 21, respectively. The 21-day total float means that task 2.2.4, with a duration of 5 days, can begin at any time within that window. The free float between two tasks is determined by subtracting the early finish of the predecessor from the early start of the subsequent task. For this example, 48 – 43 = 5 is the free float between tasks 2.2.6 and 2.2.4. Therefore, task 2.2.4 can slip 5 days, using up 24 percent (5/21) of the total float and all of the free float. If task 2.2.4 slips 6 or more days, then the total float for task 2.2.6 decreases. As stated in the first part of this chapter, there is little concern if the free float is used, but there will be a great deal of concern when the total float disappears.

There are many tasks along noncritical paths which have a great deal of total float but no free float. An example of this is task 2.1.1 as the predecessor and task 2.1.2 as the successor. Task 2.1.1 has 16 days of total float, but there is zero free float between the two. Watch carefully for these situations because there is no free float available for slippage.

8. Look for opportunities of optimization. When the schedule has been completed, you should try to identify places where the time line could be shortened or tasks might be performed in parallel. There may be tasks along the critical path where the durations can be shortened. Task 1.1.4, "Schedule survey," is the second-longest task on the critical path. Maybe the 15 days can be shortened to 10 days, but that cannot be done by the project manager alone. Referring back to the WBS worksheet, we find that the HR manager is responsible, with the department managers as resources. This task is *theirs,* and they must agree to any change.

Chapter

8

Step 6: Start the Project

Laying the Groundwork

By this point in the project management process, an agreed-upon plan exists. As a minimum, that plan consists of:

- A definitive statement of work
- A technical definition of the product or program being developed
- A plan for assuring the desired quality in the product
- A list of the tasks to be performed with the projected resources, cost, and schedule for each
- An analysis of the manpower needed
- An analysis of the cost per month
- A proposed organization for the project

Now, the effort must change from being primarily planning to being primarily management. The plan establishes the point of departure and is the management standard against which performance will be measured. Changes in the plan must be controlled, agreed upon, and documented.

This step is the start-up of the project. Like most start-ups, it has some unique activities which set it apart from the subsequent ongoing operation of the project.

- Authorization to proceed must be obtained from management.
- Commitment of resources must be obtained from the line organizations.
- Contracts must be written for services to be purchased.
- Long-lead materials and parts must be ordered and placeholder purchase orders issued.
- Leaders in the project organization must be indoctrinated and, if necessary, trained.
- Project meetings must be scheduled for the life of the project.
- Guidelines for project teams must be distributed.
- Project teams must be motivated and their leadership established.
- A baseline evaluation of the success factors of the project should be made.
- A matrix for routine report and information distribution must be made.

Authorization to proceed

The first step in the start-up of a project is to obtain management authorization to proceed. Authorization is usually in the form of a partial budget release for start-up activities. This authorization is obtained by management review of the plan and approval of it. Depending on the size and complexity of the project, this may be a very formal procedure or just a review and approval meeting.

Commitment of resources

Obtaining the commitment of resources from the line organizations is never an easy task. It will be easier to keep the discussions with the line managers in perspective if they are viewed as negotiations. The question is not whether you are a negotiator—you are, like it or not! In these negotiations, the line manager owns the resources, but the project manager owns the work and is therefore the line manager's customer. This recognition places the negotiations in a different perspective than usual and levels the playing field. So, act like a customer.

If the planning of the project was participative and the plan was approved by the involved organizations, the line manager participated in the planning of the team's work in the project. The line manager should have a copy of the task schedule for his or her people and the histogram of the personnel needed on those tasks. With this information on the table, you should be able to frame the negotiation to resemble a planning session. Compromises will have to be made on both sides, but for the best long-term results, these should be done in a cooperative spirit.

This raises the question of how we develop and maintain this cooperative spirit in negotiations with line managers. The most important idea to keep in mind during these discussions is that line managers and project leaders have different modes of management and different operational goals, as shown in Fig. 8.1. Each comes from a different part of the organization. Line managers are part of the permanent vertical organization. They manage primarily by the authority of their position. Their operational goal is to keep the efficiency of their department as high as possible. Project leaders, on the other hand, are in a temporary position which has little or no direct authority. That is, a project leader cannot direct someone to do something. Many project leaders have bemoaned that they don't have more authority to get things done. They don't have it, they are not going to get it, and it's a good thing. If they had more authority, they would only confuse the people doing the work by giving orders which are con-

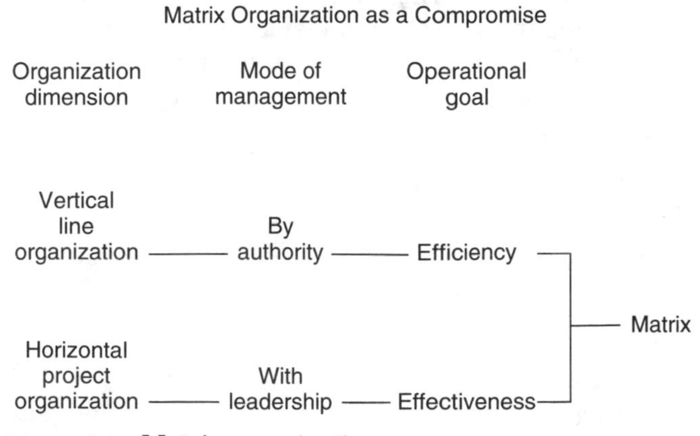

Figure 8.1 Matrix organization.

tradictory to the line manager. One reason for this inevitable contradiction is that the project leader has effectiveness as an operational goal. The project leader does not have the luxury of repeatability in a project. The work goes through only once, so there is no basis for measurement of efficiency on a project. Therefore, a project leader focuses on effectiveness—doing the job well. As a corollary to this, there is an opportunity for efficiency gains from project to project if we document performance.

The differences in management mode and operational goals between line managers and project leaders in a matrix organization make negotiations and compromises necessary. In this type of negotiation, it is critical to recognize that the only person we can change is ourselves. We cannot change the other party. In order to be effective at changing ourselves, we must first understand our options in approaching the other party, and work within those options.

In getting a better definition of our individual approach to negotiations, one author* classifies our options as follows: persuading, compelling, avoiding, collaborating, negotiating, and supporting.

By answering 45 questions, we find out our preferred approach to these negotiations. Just as important, we find out what other options we might be able to use effectively, and which ones we are likely to flub. The definitions of these approaches are:

Persuading	Convincing the other person that our viewpoint is better
Compelling	Directing someone to do something
Avoiding	Not going to get the commitment, or completely agreeing to the line manager's revised commitment of people even though it will hurt the project
Collaborating	Working as a team to make a shared commitment which will benefit the project
Negotiating	Cooperating, but retaining some vested self-interest
Supporting	Making some concession to help the line manager with a real problem

---

* Leas, S.B., *Discover Your Conflict Management Style,* Alban Institute, Washington, D.C., 1989.

There are several good training courses, books, and tapes on this type of negotiating. Some appear as footnotes in this chapter. A negotiation is the working out of a compromise to differences.

In summary, there are two principal players in the negotiation for commitment of people to the project: the project leader and the line manager. The role of the project leader is summarized in Fig. 8.2, and that of the line manager in Fig. 8.3.

Contracts for services

Most industrial projects contract for some kind of service. This can range from a few thousand dollars for temporary help to fill needed positions to multimillion-dollar contracts for design and construction of dies for manufacturing metal or plastic parts. Although the requirements are specified by people in the project team, the execution and administration of the procurement is usually performed by purchasing professionals. The legal communication to the supplier is a purchase order or formal contract. Regardless of the scope and cost of the service, there is a procurement process which must be managed to avoid misunderstandings, errors, delays, added cost, or even litigation.

The Project Management Institute* identifies the following procurement process:

1. Decide on an acquisition method.
2. Obtain authorization and funding.
3. Decide on the appropriate contract type.
4. Prepare the procurement documents with performance and financial control procedures included.
5. Invite qualified suppliers to bid.
6. Receive responses.
7. Evaluate responses.

---

* *Project Management Body of Knowledge,* Project Management Institute, Drexel Hill, Pa., 1987.

Project Leader's Responsibilities

- Sequentially plan and integrate the work flow.
- Coordinate the application of resources to the work.
- Lead and guide the project team.
- Integrate into the company's management system.
- Track and document the project's progress.
- Prepare and present milestone reports to management.

Figure 8.2 Role of the project leader.

Line Managers' Responsibilities

- Maintain control over the orderly use of your department's people and other resources.
- Recognize that you represent the operational environment for projects.
- Recognize that projects are fragile things which can be killed easily, even by neglect.
- Negotiate in good faith for the scheduling of your people and other resources.
- Participate actively in the effort to solve resource problems as well as technical and cost problems.
- Recognize that the project team leaders help you organize and control the work your department performs.

Figure 8.3 Roles of line managers in a matrix organization.

8. Select a supplier.
9. Issue purchase order or negotiate contract.
10. Cope with bid complaints and protests.
11. Manage supplier's performance and cost.
12. Manage changes in specifications and requirements.
13. Settle contract disputes.
14. Close contract.
15. Evaluate procurement and supplier performance.

These steps are not mutually exclusive. They overlap into an integrated procedure in which parts of several span across steps. This list of steps provides a checklist for planning and managing a procurement.

Decide on the acquisition method. There are several ways that services can be acquired:

- By advertising for suppliers
- By inviting selected suppliers to bid
- By negotiating with a sole source
- By writing a purchase order to a known supplier

Obtain authorization and funding. This comes from the company management's review of the project plan as described previously in this chapter.

Decide on the appropriate contract type

Unit price	A fixed price per unit with final price dependent on the quantity
Firm fixed price	A lump sum contract
Fixed price plus incentive	A fixed price for specified performance plus a predetermined fee for superior performance
Cost plus fixed fee	Reimbursement for allowable cost plus a fixed fee which is paid incrementally as the project progresses
Cost plus percentage of cost	Reimbursement for allowable cost plus an agreed-upon percentage of cost as profit
Cost plus incentive fee	Reimbursement for allowable cost plus a predetermined fee for superior performance

Fixed price contracts are used where there is a relatively high degree of certainty in performance and delivery. It places the risk on the supplier. Cost-type contracts are used where there is significant uncertainty in performance and delivery. The variations of cost contracts are intended to provide risk sharing between the supplier and the purchaser.

Prepare procurement documents with performance and financial control procedures included

- Technical specifications
- General requirements

  Delivery
  Schedule controls
  Cost controls
  Quality controls
  Management of technical changes
- General terms and conditions
- Supplementary terms and conditions

Invite qualified suppliers to bid

- Identify potential suppliers.
- Conduct prequalification reviews of

  Experience
  Past performance
  Capabilities
  Resources
  Current work load
- Verify qualification data.
- Rank prospective suppliers by their qualifications to provide the service.
- Develop a bid list of qualified suppliers.
- Hold a preselection meeting to clarify the requirements before bids are submitted.
- Invite qualified suppliers to bid.

Receive responses. *Caution:* There is sometimes an ethical breakdown at this point because of the competitiveness of suppliers and the value of the contract. Try to prevent it.

- All bids or proposals should be due simultaneously.
- There should be no collusion on the part of the suppliers.

- There should be no release of information from the purchaser until all bids and proposals are evaluated.
- After bids or proposals are evaluated, the proprietary rights of suppliers should be honored.

Evaluate responses. Rank responses in terms of

- Technical merit
- Quality assurance
- Estimated total cost
- Delivery expectations
- Technical and financial risk
- Other stated considerations

Select a supplier. Select the supplier who best matches the service needed and the limitations of the procurement.

Issue a purchase order or negotiate contract. Negotiation with a supplier is similar to negotiation with a line manager for commitment of resources. The primary difference is in the contractual arrangement with the supplier.

Cope with bid complaints and protests. No one wants this to happen, but occasionally it does. The cause is usually traceable to an unethical practice in the reception of responses, evaluation, or award. However, an unusual practice or an unexpected practice may trigger complaints. They are best dealt with by investigation and full disclosure of findings to all parties involved. Serious protests may involve litigation.

Manage supplier's performance and cost. Traditionally, suppliers are given a contract and turned loose until delivery day. It is a better practice to integrate the supplier into the project plan and schedule. In this way, the supplier is treated like a full partner in the project just like an internal department. The same kind of plans are expected and the data collection cycle applies to the supplier just as it does to all other organizations.

Manage changes in specifications and requirements. This is an area which cannot receive too much attention. It is the source of more havoc in the relationship between suppliers and projects than any other problem. Especially in product development work, changes are inevitable. But their timing and impact can be managed. In fact, changes need more project management than in-line development work. Suppliers are especially vulnerable to missed communications or omitted transmission of change information because they are not continuously present. Both the supplier and the project team should guard against making assumptions about what the other party knows or is doing.

Settle contract disputes. Serious contract disputes should not happen in industrial projects. They are the result of misunderstandings and erroneous assumptions. Intelligent project planning and alert project management will eliminate almost all of the causes for contract disputes. The planning should be realistic, have the participation of those involved, and be agreed to by all involved parties. Erroneous assumptions are avoided by asking a lot of questions and documenting changes and change orders and their impact. When a dispute arises, the struggle will be to keep it from developing into litigation.

Close contract. Contracts are usually closed at the completion of the work. The supplier receives a final payment and an acknowledgment of the contribution to the project. The other closing, which is not as nice, is when the supplier defaults on the contract. This situation is usually handled by the purchasing professionals because of the tendency for the situation to go into litigation.

Evaluate procurement and supplier performance. It is important to subsequent projects for the things which could be improved in the procurement process to be documented. There is always room for improvement, especially in the relationship with suppliers. Try to be objective about whether any deficiencies in the supplier's performance might have been caused by deficient practices in the project.

Ordering long-lead-time items

The procurement of long-lead items has some special considerations. The long lead time can be from six months to a year and a half in industrial projects. Machine tooling for high-volume production can take the longest time from the order to the time when it is production-ready. This means that frequently in product development projects, we must order the tooling before we have a product design to work from. In this situation, a supplier is selected very early, brought into the design team, and given a contract to participate. This allows the tooling supplier to provide input into the product design relative to the tooling, and to be close to the product design for preliminary tooling design and planning. The tooling company includes this work in their future manufacturing schedule.

Design changes and the communication of design changes are the biggest risks to long-lead items. Because of the long lead time, it is almost disastrous for a machine or tool to arrive which no longer produces the current part. Nevertheless, it still happens, and only extra project management effort will prevent it.

Training of team leaders

All leaders of teams on the project need an orientation session on the objectives and strategy of the project and their role in it. Those who lack experience in project management also need training in the fundamentals of project management. Then they need experienced advice as the project progresses.

Scheduling of project meetings

After a while, the priority of project meetings tends to drop almost out of sight, and attendance diminishes almost to the point of being ineffective. There are two things which can be done to keep the priority of project meetings near the original level of importance. First, establish a recurring time slot for the meeting. This puts it on the calendar of all participants for the life of the project. Second, devote the meetings to group problem solving, at least to identifying problems, discussing options and limitations, and determining who has responsibility for solving problems and in what time frame. Meetings should not waste time on project-status reporting. That can be accomplished better in other ways. Make project meetings valuable time.

Distributing project guidelines

This may include a company's blanket guideline, but procedures specific to the project should also be distributed. Distribution should be to all participants. We should never make the mistake of assuming that everyone understands the procedures we are going to use.

Motivating project teams

Project teams have a life of their own, but some never come to life. We breathe life into a project team by motivating its members to become active participants. One of the greatest motivators is for people to see that they are making a visible contribution. Feedback is what makes the contribution visible. We should never hesitate to point out to people how they have contributed to the project. It keeps morale high, even when the going gets tough.

Making a baseline evaluation of the project's success factors

Everyone has an opinion about the chances of success for the project when it first starts, but none of these are objective or broadly based. A better method is needed, and one exists. It was developed by Slevin and Pinto, and is presented in their article, "The Project Implementation Profile."*

It is a measuring instrument for determining the relative strengths and weaknesses of a project. When completed, it provides a profile of ratings for 10 key success factors of the project. When these are compared to the standard profile, it provides a rating of the probable success of the project. Thus, we can determine the weaknesses of our project and whether those weaknesses are life-threatening.

The 10 factors measured by the profile are:

- Communication
- Project mission

---

* Slevin, D.P., and Pinto, J.K., "The Project Implementation Profile," *Project Management Journal,* September 1986, pp. 57–65.

- Top-management support
- Project schedule/plan
- Client consultation
- Personnel
- Technical tasks
- Client acceptance
- Monitoring and feedback
- Troubleshooting

Each of these 10 factors is evaluated with 10 questions. The questions probe the extent to which we are satisfying each factor. These are questions which a project owner and team leaders should be asking anyway. As many people on the project team as possible should contribute to the answers. Discussion of disagreements will bring latent information to the table as well as help build consensus and cohesiveness in the team.

The results of analysis of the profile will show project strengths and weaknesses as a standardized percentile rank for each of the factors. This can be displayed graphically, which makes it suitable for distribution to the project team and to management. Some weak factors the team can improve upon may take management involvement through the project sponsor. It helps keep the project visible.

Project management is a complex process and the profile gives us a measure of the health of the entire project. Taking this measurement at start-up gives us a baseline measurement. It can be repeated at designated points in the project to monitor the strengthening of weaknesses and to determine whether any new weaknesses have developed.

Report and information matrix

At start-up, we need to plan who gets which project reports. Managers prefer reports which are specific to their interests. Executives want summaries. The project team members want to know where they are and to be reminded what they need to do next.

All these different information requirements indicate that we need a method of organizing the production and distribution of information. This is done by developing a list of the reports available from the database, having each manager or organization select their preference, and arranging this distribution information in a matrix which tells us who gets what. A sample of a report distribution matrix is shown in Fig. 8.4.

Starting the TQM Implementation

The planning functions of the TQM project implementation have been completed. Steps 1 through 6 of the project management process represent both management and project planning. Step 7, starting the TQM project, is next.

As you have seen in Chap. 8, there are 10 activities which constitute a project start-up. We will examine most of those activities and apply them to the case study which has been developed. Since there has not really been a "start-up," examples will be given of potential scenarios for the case. The examples will not be all inclusive, so the reader is again challenged to search for other possibilities.

Authorization to proceed

In our example of the TQM implementation, the authorization to proceed would most likely be informal. Although the communications portion of the project is companywide, the statistical methods apply only to the manufacturing areas.

Report Distribution Matrix

Report description	Project manager	Project office	Team member	Line manager	Executive management
Monthly summaries	X	X			X
Milestone reports	X	X	X	X	X
Personnel histograms	X	X		X	
Pressure test report	X	X	X		
Humidity test	X	X	X		
Issues report	X	X	X	X	X
Scheduling summaries	X	X	X	X	

Figure 8.4 Report matrix.

As a show of top-management commitment to quality, the president of the company would probably issue a letter to the project sponsor, project owner, and the highest level of line management. The responsibility for communicating project authorization to the lower levels of the company rests with line management.

Commitment of resources

The beginning of Chap. 8 defines how the project leader can go about negotiating with the line manager for resources. Our project owner, C. P. Kaye, must deal with the following line functions for his resources:

- Quality
- Human resources
- Purchasing
- Manufacturing

The level of effort required to negotiate with each of these line functions will vary within the company, and from company to company. The quality department, for instance, may be so pleased to have TQM started after years of trying to convince top management that they will be supercooperative. On the other hand, quality departments always seem to be understaffed. In this case, it may be difficult to negotiate resources. A closer look at the WBS worksheet reveals that, although quality is responsible for a number of tasks, the resources for those tasks are not directly reporting to the quality function. The line function will also need to be skilled at negotiating for resource time of a particular task.

Contracts for services

Although the TQM project described here is relatively small, there are several tasks which will require some form of contract. The consultant, the printer, and the communications media all require contracts. The printing of the surveys is probably the least complex, and requires only a purchase order. The communications media are very uncertain at this point, but have the potential for being complex. Hiring a consultant may initially appear to be an easy task. Finding one who has the knowledge and experience and who fits into your organization is sometimes difficult. The following is a summary for each contracted service.

Printing

Acquisition method: Bid invitations or purchase orders to known supplier

Contract type: Unit price

Communications media

Acquisition method: Bid invitations or sole-source negotiation

Contract type: Firm fixed price

Consultant

Acquisition method: Bid invitations or sole-source negotiation P.O. to known supplier

Contract type
Firm fixed price
Hourly rate

Remember, when purchasing any product or service, it is important to evaluate all aspects of the potential supplier. They are part of *the business system,* and must fit in with your company's goals and strategy in addition to cost considerations.

Scheduling project meetings

As discussed in this chapter, the project meetings to be scheduled here are the down-and-dirty, in-the-trenches, problem-solving meetings. Having a meeting to status report is a complete waste of time at this level. Our TQM implementation project is to be completed in 84 days, which is not much time in the first place. We would want to establish a project meeting schedule for hiring the consultant, establishing the communications media, and implementing the statistical methods. The latter two correspond to the actual project phases, while hiring the consultant is a task important enough with a long enough duration to warrant project meetings. In all three of these cases, the recommendation would be for weekly meetings to keep the project tightly on track.

Project guidelines

Guidelines need to be established for the project in terms of reporting, responsibility, meetings, etc. Again, this is a somewhat small TQM implementation, so the guidelines can be brief. For instance, if something affects a task which impacts, or may impact, the project schedule, there should be a reporting procedure to follow. Waiting for the next scheduled meeting could mean disaster. Included in the procedure might be provisions for vacation, sick leave, or other unplanned absences by key personnel.

An example of a reporting procedure could be:

> All changes that impact the project schedule must be reported within 24 hours to the project leader. If the project leader is not available, then the project owner and the project sponsor, in that order, must be notified.

Other procedures might include seemingly trivial matters, but they are important to the project's success.

> The communications project team will meet every Thursday at 1:30 p.m. in conference room A. All members are expected to be there ***on time!*** Project team members late to the meeting will be required to contribute $.50 per minute to the "tardy pool." At the end of the project, tardy pool proceeds will be given to the team member(s) with the best on-time record. Excused absences can be granted in advance by the project leader.

Who said the project had to be boring?

Report and information matrix

It is important to define what reports will be generated and who will receive them. The reports most important to the TQM implementation would be:

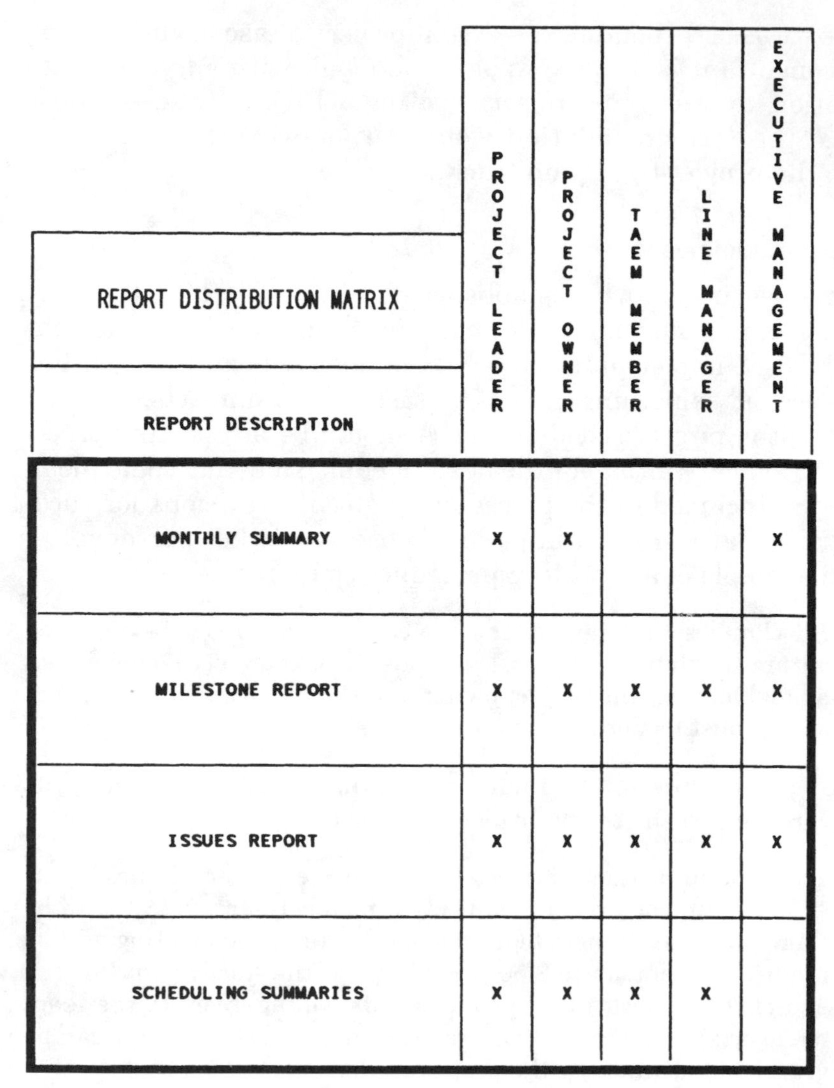

REPORT DISTRIBUTION MATRIX

REPORT DESCRIPTION	PROJECT LEADER	PROJECT OWNER	TAEM MEMBER	LINE MANAGER	EXECUTIVE MANAGEMENT
MONTHLY SUMMARY	X	X			X
MILESTONE REPORT	X	X	X	X	X
ISSUES REPORT	X	X	X	X	X
SCHEDULING SUMMARIES	X	X	X	X	

Figure 8.5 TQM report matrix.

1. Monthly summary
2. Milestone report
3. Issues report
4. Scheduling summaries

The report matrix, Fig. 8.5, shows the distribution to the various functions.

Chapter

9

Step 7: Track Progress and Identify Problems

The Project Management Toolbox

Now that work has started, we are faced with the prospect of having to actively manage the accomplishment of tasks. Please note that we are not managing the work within the activities. The people are trained, experienced, and competent. This is true for a vast majority of tasks, so we empower people to do their jobs.

Problems almost always arise because of mismanagement. Personnel were not told correctly what to do, they were not directed at all, conflicting instructions were given, the deliverable result was not delivered, the poor quality of the deliverable was ignored, urgency was not communicated. None of these have to do with supervising the work directly. There is a message for us in the anonymously penned cartoon of Fig. 9.1. If the coxswain tries to direct the stroking, chaos results and the boat probably will end up pointed in the wrong direction. The coxswain's job is to maintain the rhythm and keep the boat pointed in the proper direction. For a project leader to maintain the rhythm, he or she must track the myriad of handoffs between tasks, ensure that they happen when they are supposed to happen, and confirm that the quality is as expected. The project leader has several tools in the project management toolbox to help with this: a closed-loop information system, project meetings, problem solving, and reporting.

Figure 9.1 Managing effectively in a project environment.

Closed-loop information system

When we begin to use the project plan to manage the project and track progress, we have a powerful tool in the idea of a closed-loop information system. This is not a structural tool. It is a procedural tool which we must make work. It principally involves getting task information from the people doing the work, then feeding back to them the status of the overall project. This is the principal information channel in project management. For information in the project, the people doing the work are our primary customer. Reporting to management is a secondary obligation.

This loop contains the four traditional information activities: collecting data, analyzing data, interpreting and problem solving, and communicating the overall status, as shown in Fig. 9.2.

This cycle should be completed in as short a time as practical. This involves consideration of how long it takes to show progress, the duration units in the scheduling model, and how quickly the information can be collected and processed. The most common project information cycle times are one and two weeks. Some projects cycle their progress information monthly, but they are multiyear projects with slow progress rates. Only in an emergency do projects cycle their information daily or hourly. A sample schedule for a two-week project information cycle is shown in Fig. 9.3.

Collecting task data. The information cycle is driven by collection of task data. We all want data on the project, but few of us are willing to collect it, especially on a regular basis. It must be done or we will lose control of the project.

This is a process of interviewing. It is most successfully done by going to the people doing the work and asking about the tasks in work. It is usually unsuccessful to ask, "How is it going?" The routine answer is, "OK." A more complete and specific answer is

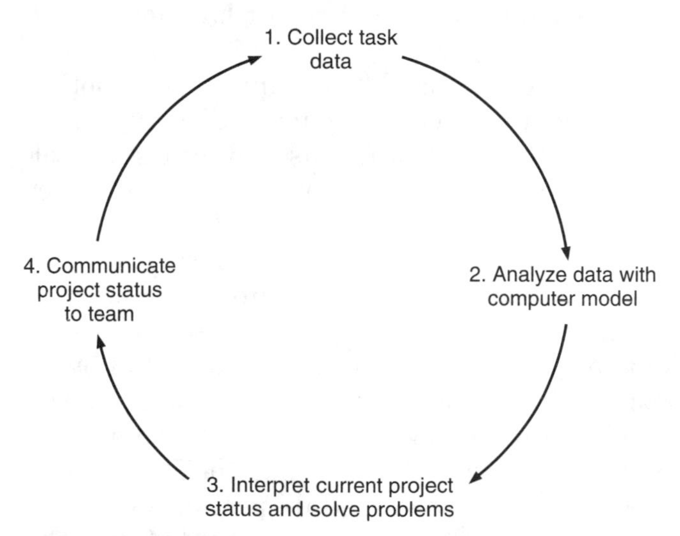

Figure 9.2 Closed-loop information system.

Development Project Update Cycle	
Week 1	
Tuesday, Wednesday	Get markups from suppliers, clarify problems
Thursday, Friday	Enter updates into models and reschedule
Week 2	
Monday	Sort and make schedule reports for suppliers
Tuesday, Wednesday	Print and deliver updated schedule reports; review with project engineer
Thursday–Tuesday	Suppliers review and update schedules
Tuesday, Wednesday	Get markups from suppliers; clarify problems

Figure 9.3 Two-week project information cycle.

needed. A specific question should be asked about the task being performed. It should be an open-ended question which takes more than a simple yes or no to answer. When things are not going OK, the person reporting must trust the interviewer not to use the information to hurt him or her. To a great extent, the fidelity of a response in less-than-favorable circumstances will depend on whether the company culture and project culture are problem-solving-oriented. Everything should be done to reduce the anxiety of people when they have to report bad news. And, please, don't shoot the messenger.

People are always more comfortable talking about another person's problems, or about problems they may be having which are caused by others. In this way, problems and issues are collected at the same time schedule status data is collected. Clues to problems of others should be followed up with the involved person to verify them and obtain more specific information.

If sufficient personal contact has been built up, collection may be done by telephone. But it should be interspersed with personal contact. Some project leaders expect task data to flow to them from the people doing the work. This just doesn't happen. Updating is an administrative task, and even well-intentioned people can get too busy to provide the information. This is especially true when people are under schedule pressure or when things are going wrong. Automatic reporting tends to fail at the very time accurate updates are needed the most.

When people or groups are reluctant to report their progress, and don't report, an interesting thing happens. Because their status data shows no progress during the period, their tasks are likely to be highlighted on the computer output as being on the critical path. This brings them under closer scrutiny from management. So, there is an incentive for people to cooperate and provide progress information on time.

Analyzing the data with the computer model. This step involves entering the data into the project model in the computer and running it. The output reports provide new project status information. If the resultant status forecasts schedule slippage or personnel overloads, it needs to be highlighted to the project team at the meeting. This will require some kind of corrective action which must be agreed upon by the team.

Interpreting current project status and solving problems. Interpretation of current status requires that the current project status be compared to the planned status at the present time. The planned status is usually contained in the baseline model which was the result of front-end planning. That baseline may have been revised subsequently by agreement. The most frequent cause of baseline revision is the recognition of circumstances which were not originally anticipated, such as a serious underestimation of a task duration or a completely missed set of tasks. Changes in the baseline plan must be agreed upon by all involved (signed off) and documented as to what the change was, what caused it, and when it was made.

When schedule slippage for completion of a task causes downstream schedule slippage or personnel overloads, corrective action must be taken. The purpose of the corrective action is to bring the project back on schedule or within its personnel limits. This forward projection is one of the unique advantages of using a computerized logic network project management model in decision making. It also provides a single source of information for all the people who must participate in solving the problem.

When changes are made in the project because of problem solving, these changes should be input to the project model and rerun. These runs may be tentative and experimental to determine how much of a solution each represents. For these experimental runs, copies of the current model should be used so the current situation is not lost. The copies can be used experimentally to simulate the results of these trial corrective actions. When the simulation of a corrective action produces an acceptable situation, that model should be proposed as the new plan. It will have to be agreed to by all participants. When approved for use, this version must be coded to make it clear that it supersedes the previous version of the project plan, and it must be distributed to all participants.

All problems and issues in the project must be documented in a "problem and issue log." A sample log is shown in Fig. 9.4. Each problem or issue should be numbered with a code which ties it to the task to which it is related. This facilitates including problem-solution time in the project schedule updates. That way, project updates include problem status as well as task status. The "subject" column is for identification of the task or part involved in the problem. The "responsibility" column names the people or group working on the problem. The "accountable" column names the person to contact for information (i.e., the one who is managing the problem solution work). Entries in the log are shaded to indicate that the problem has been solved. Closed problems and issues are not deleted, so at the end of the project, an analysis of things gone wrong can be done. Problem-solving considerations and techniques are covered later in this chapter.

Communicating project status to the team. After the interpretation and solution of problems, the resulting plan must be made available to all participants so it can be used as the new project plan. Then, updates will be made to this new plan. It must be delivered to the participants with sufficient time to allow them to review it and update their status on it.

Communication of project status information is done by use of versions of reports from the project management software. The specific format of these reports will vary, depending on the software used.

The most common status reporting format is the Gantt chart. It is a bar chart with a bar representing the active time for each task listed, as shown in Fig. 9.5. Each task is identified with its WBS code to avoid ambiguity where descriptions are similar. The duration of each task is shown, along with the currently

No.	Date opened	Subject	Description	Responsibility	Accountable	Status	Solution duration	Date closed
130	2/25/92	Rear bumper attachment	Confirm that the rear bumper lower attaching tabs are at the correct angle. (First sample did not appear to line up.)	G. Cogdon B. Burns D. Balynt	R. Brose	In progress	11/17/92	
132	3/10/92	Rear facia/ Floorpan rat hole	At EP-build resolve any potential "rat hole" at floorpan/facia interface. Estimate *cost* vs. *customer want.*	G. Cogdon B. Burns D. Balynt	R. Brose	In progress	11/17/92	
140	3/24/92	FR and RR bumper TPO	Resolve part differences.	T. Dellock L. Byrd	R. Brose	In progress	11/17/92	
151	9/8/92	Rear bumper facia attachment	Part #17A881 Revise to make easier to install.	G. Cogdon B. Burns L. Byrd	R. Brose	In progress	11/17/92	
152	9/8/92	Rear bumper facia attachment	Revise to eliminate exhaust pipe interface.	G. Cogdon B. Burns L. Byrd	R. Brose	In progress	11/17/92	
153	9/8/92	Rear bumper facia	Add stop pad to rear half of top of facia; color-keyed, adhesive attachment, friction surface, and 5 MPH proveout.	G. Cogdon B. Burns L. Byrd	R. Brose	In progress	11/17/92	11/17/92
154	9/8/92	Rear bumper facia	Add clear urethane to rear underside of facia to prevent paint chipping from Stone Pecking test. Supplier formal quote is due.	G. Cogdon B. Farriss L. Byrd	R. Brose	In progress	11/17/92	11/17/92
159	11/17/92	Rear bumper facia mold in color	Revise the rear bumper to provide mold in color.	G. Cogdon B. Farriss L. Byrd	R. Brose	In progress	11/17/92	

Figure 9.4 Problem-and-issue log

Report: BUMP2 | Data date: 02OCT92 | Page: 3
Project: Bumper | Date: 05OCT92 | Time: 15:55:21

Legend: In progress; Planned; Critical; Baseline; Complete

Task ID	Description	Duration	Current Start	Current Finish	TF
BFB263	FRT BPR beam procure press	200	01JAN92	01MAR93	2
BFB246	FRT BPR beam die set, develop spec	40	02MAR92	01OCT92	5
BFB291	FRT BPR beam design checking fixt	20	06MAY92	30OCT92	74
BFB247	FRT BPR beam assembly welder, develop spec	40	06MAY92	01OCT92	10
BFB248	FRT BPR beam strap, develop die spec	40	06MAY92	01OCT92	105
BFB230	FRT beam design die set	30	08SEP92	29OCT92	5
BFB232	FRT BPR beam design assembly welder	20	02OCT92	29OCT92	10
BFB249	FRT BPR beam design strap dies	10	02OCT92	15OCT92	105
BFB282	FRT BPR beam final FMEA & control plan	10	02OCT92	15OCT92	–22
BFB258	FRT BPR beam & strap CAD file	5	02OCT92	08OCT92	30
BFB259	FRT bumper beam planja modify dies set	10	09OCT92	22OCT92	30
BFB239	FRT BPR beam construct strap dies	50	16OCT92	06JAN93	105
BFB251	FRT beam make VP parts	10	23OCT92	05NOV92	30
BFB236	FRT BPR beam construct die set	150	30OCT92	09JUN93	5
BFB242	FRT BPR beam const assembly welder	120	30OCT92	28APR93	10
BFB292	FRT BPR beam check fixt construction	80	02NOV92	04MAR93	74
BFB265	FRT beam PIST/PIPC VP parts	10	06NOV92	19NOV92	30
BFB241	FRT beam strap die tryout	5	07JAN93	13JAN93	105
BFB264	FRT BPR beam install press	40	02MAR93	26APR93	2
BFB267	FRT BPR beam install furnance, descalers & auto	20	27APR93	24MAY93	2
BFB244	FRT beam install assembly welder	20	29APR93	26MAY93	10
BFB268	FRT BPR beam tryout press line	15	25MAY93	14JUN93	2
BFB269	FRT beam modified ISM	60	27MAY93	18AUG93	10
BFB253	FRT beam make 1PP parts (planja)	20	27MAY93	23JUN93	10
BFB240	FRT beam install/try out die set	15	15JUN93	05JUL93	2
BFB254	FRT BM PIST/PIPC 1PP parts	10	24JUN93	07JUL93	40
BFB260	FRT BPR beam ISM	30	06JUL93	16AUG93	2
BFB252	FRT BPR beam manufacture 4P parts	20	17AUG93	13SEP93	2
BFB266	FRT BPR beam PIST/PIPC 4P parts	10	14SEP93	27SEP93	2
BFB315	FRT BPR beam & isolator start maintain drawings	0	02FEB94	01FEB94	144

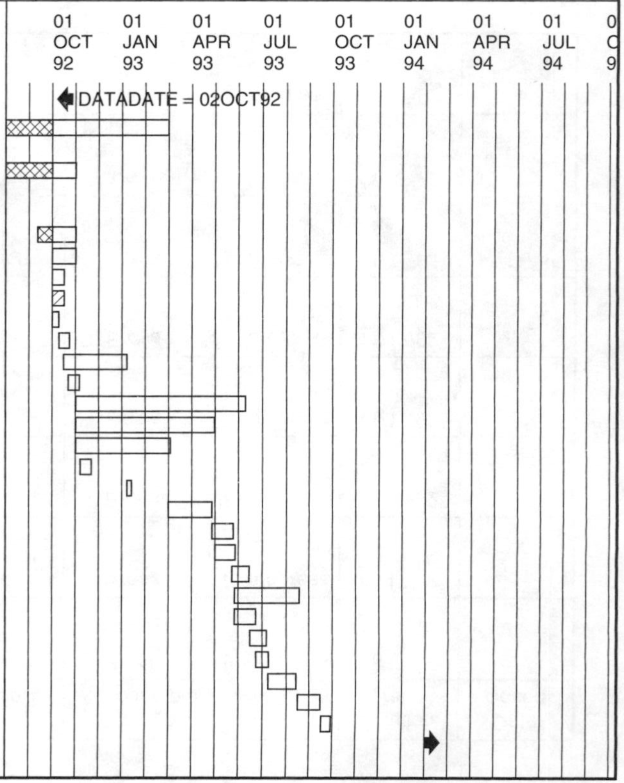

Figure 9.5 Sample Gantt chart.

scheduled start and finish dates. The bar in the calendar field starts on the start date and ends on the finish date. The bar is hollow to show the planned active time, and shaded for work completed. Milestone dates are shown as diamonds.

Figure 9.6 shows a variation of the Gantt chart. It has two displays of bars for each task. One set of bars is for the entire length of the project. The other set of bars, shown to the left of the tasks, is for a period of only four weeks. This technique is called *windowing*. It is used for focusing attention on currently active tasks in long projects.

Windowing allows a project leader to manage the collection of progress data with only three questions:

1. For those activities which were to start since the last update: "Did it start?"
2. For those tasks which were to finish since the last update: "Did it finish?"
3. For those tasks which are ongoing during the period since the last update: "When will it be finished?"

These three questions are no-nonsense. The first two require a yes or no answer, no evasiveness. If the answer is yes, it's OK. If the answer is no, then an explanation is requested. The third question requests a verification of the planned completion date or a proposed new one. If the date is verified, it's OK. If a new date is proposed, an explanation is requested. The severity of the schedule change may require replanning the entire project with involvement of the team. Before long, everyone will recognize the seriousness of completing their tasks on time.

Please note in the preceding example that we did not ask what percent of the ongoing task was complete. Experience has shown that that question does not get an accurate answer. In project management, for many years it has been repeated that when we ask for percent complete, the answer is almost always 80 percent. The remaining 20 percent of the work may take more time than the original 80 percent!

Another common format in project management is a tabular report, which will have a format provided by the software. Tabular reports are used mostly for detail information for activity teams. Few people will be interested in wading through such

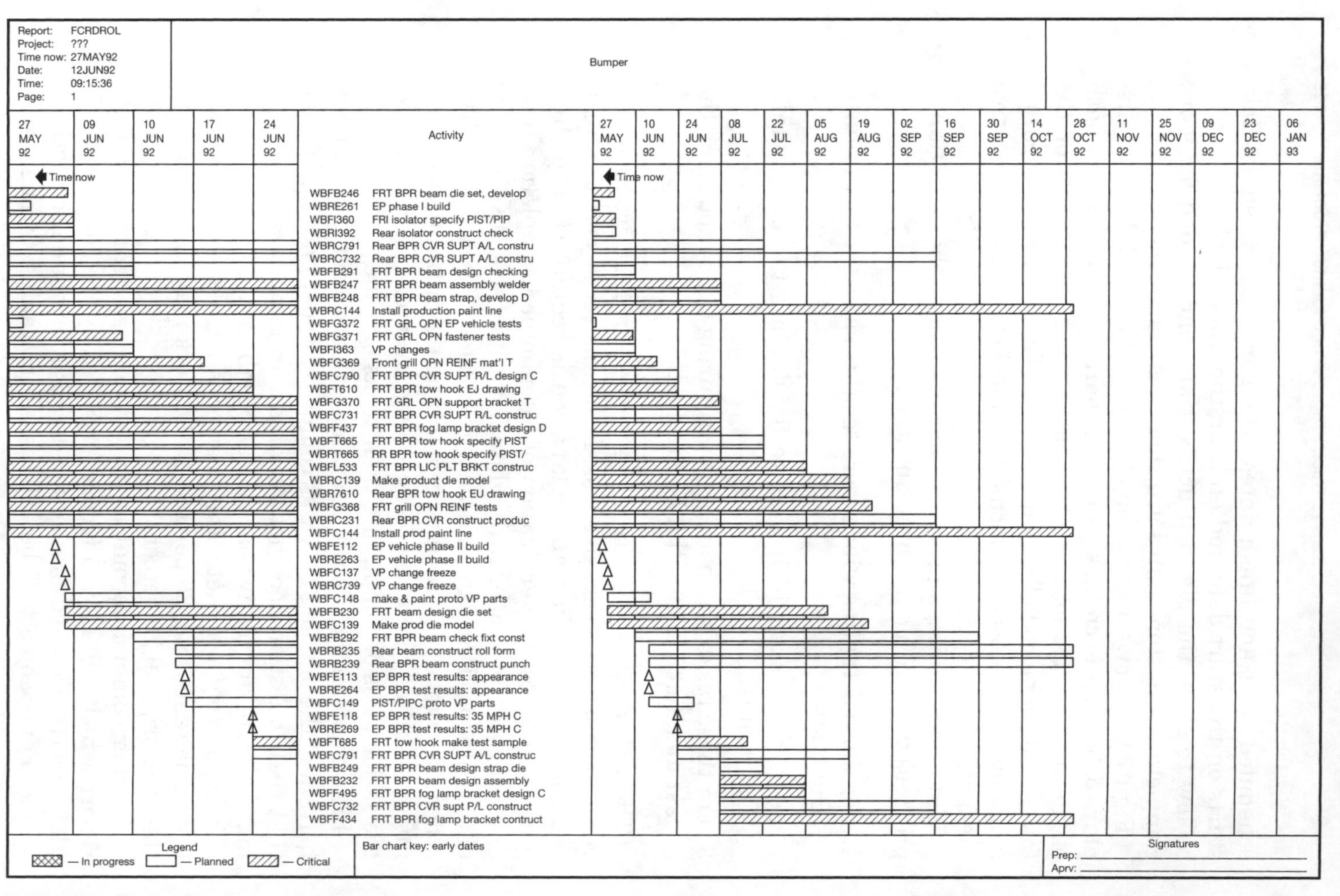

Figure 9.6 Gantt chart variation.

detail information unless it is relevant to them. This detail information may be necessary for problem solving.

The logic network diagrams are sometimes used as status reports. The logic network diagram is not intended as a status report. It is a working document of the project team. It is a document which cannot simply be read—it must be studied. Therefore, never put it in front of someone who is not working with it.

Project meetings

One of the strongest tools used to keep the work on a project coordinated is the project meeting. There are a lot of things a project meeting can be used for. Some meeting activities are more effective than others.

One of the poorest uses of project meetings is to collect project status, although it is frequently used as an administrative convenience. Even though the situation is face-to-face, it is not one-on-one. People on a project consider status to be far less important than issues and problems. Meetings which are used to collect status are eventually poorly attended. There is a better way to collect status information.

Project meetings can be used to report overall project status to everyone. This includes the impact of problems and other issues. Such feedback meetings have value to people, so they will attend.

The best use of project meetings is to hold group discussions of problems and issues. This is not limited to defining the problem, but includes assigning responsibility and estimating a completion time. If it takes a team of people from several departments, arrangements for the establishment of this problem-solving team should be made. This kind of meeting serves to reinforce teamwork and coordinate group efforts.

Leadership. Leadership is dealt with here because the effects of leadership are at their greatest when the work is being done. Also, project meetings are the place where leadership is most visible. The leader's style and skill are on public display in meetings, so leadership behavior should be at its best.

The rationale for using leadership instead of authoritative management behavior is as follows:

- Project leaders are seldom given any formal authority, so they can neither direct people nor control them.
- Project leaders, therefore, must substitute leadership for authoritative control methods.
- Leadership is a human activity, not an administrative procedure.

Leadership has not yet been scientifically defined. It has been studied historically for centuries, and scientifically for four decades, but it is unlikely to be scientifically defined until some time around the middle of the twenty-first century.*

The first principle of leadership is that leaders must behave supportively. Support is projected by listening attentively, making supportive comments, applauding to show approval, and expressing criticism constructively. More specifically, there are two goals for the leader's supportive behavior: task-oriented support and team-cohesion support. Specific behaviors for each of these goals are as follows:

Task-oriented behaviors

Initiate or contribute. Make suggestions, propose goals, present new ideas or new ways of looking at a team problem, or suggest a process for solving a problem.

Seek information or an opinion

Ask for information and facts which are pertinent to the problem.

Seek other's ideas and suggestions.

Seek an estimation of value of problems and solutions.

Ask for feelings.

Give information or opinion

Offer facts.

Offer generalizations from personal experience.

Make suggestions.

---

* Hersey, P., and Blewshard, K.H., *Management of Organizational Behavior,* Prentice-Hall, Englewood Cliffs, N.J., 1988.

Interpret issues in terms of pertinence to team.

Pull ideas together.

Coordinate subteams.

Clarify confusion.

Suggest alternatives and develop implications.

Orient and evaluate

Summarize where the team is relative to its goals.

Summarize what has occurred.

Question the direction the team is going.

Question logic, methods, policies.

Make a record. Document problems, decisions, solutions, and suggestions to serve as the team's memory.

Take consensus

Ask the team how close they are to a decision on a scale of 1 to 10.

Poll the team members on their agreement with the decision and ask for their understanding of their responsibility for the results.

Team-cohesion behaviors:

Encourage

Acknowledge contribution of others by agreement or praise.

Show warm interest in others.

Show acceptance of a person's contribution with a smile and body language.

Harmonize the work of the team

Ask members to accept their technical, cultural, and personality differences.

Mediate disruptive conflicts.

Compromise your position.

Change your position in light of more information.

Admit error.

Be a gatekeeper

Keep communication channels open with other parts of the organization.

Encourage participation in discussions.

Encourage the sharing of opinions and viewpoints.

Reduce tension

Use humor when appropriate.

Call a break when needed.

Watch the team

Help the team evaluate its progress and process.

Comment on aspects of its progress.

Team cohesion can be damaged by the following behaviors. Watch for them.

Aggression

Putting down others

Joking aggressively

Disapproving of team values

Blocking

Stubborn resistance

Opposing beyond reason

Reintroducing issues after the team has resolved them

Seeking personal recognition

Calling attention to oneself by boasting about personal achievements

Continuing to hold the team's attention disproportionately

Self-confessing. Using the team as an audience to express opinions and feelings which are not relevant to the topic or team.

Lack of focus. Displaying a lack of involvement in the team process by being cynical, nonchalant, or prone to horseplay.

Dominating

Manipulating others on the team with flattery or coercion

Resisting team building by the leader

Repeated help-seeking ("poor me!"). Trying to get sympathy for insecurity or confusion rather than assistance, the "poor me" syndrome.

Withdrawing. Trying to remove the source of discomfort by refusing to participate and by displaying a general lack of interest

Nitpicking. Emphasizing insignificant details.

The standard agenda. Communications are clarified by following a standard agenda.

This is a practice which has been in use since the early part of the twentieth century. A general process for a project meeting standard agenda is as follows:

1. State the purpose of the meeting.
2. Define the problem as known.
3. Collect the available facts.
4. Agree on criteria for a solution.
5. Explore alternative solutions.
6. Test the alternative solutions against the criteria.
7. Refine the selected solution.
8. Document and publish the problem and selected solution.

There are several behaviors which are critical to the success of this process:

- The members must understand the purpose and goal.
- The members must be willing to interact constructively.
- The members must be willing to acknowledge other viewpoints.
- The leader must moderate the discussion.

Responses to discussion. There are several appropriate responses from the leader during a discussion. These will encourage participation and build team cohesion.

- Speak from a base of fact. If you don't know, listen.
- Active listening is participating.
- When responding to information, make a statement to clarify it, ask a question about it, or remain silent.
- Be open in your communications so others can be comfortable in adjusting their positions.
- Keep your responses appropriate to the discussion.
- When disagreeing:

  Do not interrupt the speaker.

  Introduce your objection as a fact, not an opinion.

  Do not escalate a conflict.

  Do not retreat from a conflict.

Problem solving

Problems are a part of life in projects. In fact, projects are created to solve problems. In the context of a project, problem solving drives decision making. Therefore, the quality of the problem-solving effort will limit the quality of the decision making. The worst thing that can happen in a project is for the leader to presume knowledge of the solution to a problem without even the slightest analysis of the problem. This situation is usually characterized by the exclamation, "I'll tell you what's wrong . . . so you should . . ."

Decision making in a project usually extends beyond the project realm, so problems must be analyzed carefully to be sure that the solution to a particular problem is appropriate to the total situation. Problems solved with too narrow a solution are common. These solutions either fail or are rejected outside the project team. To assure ourselves of a viable solution to a problem, we must follow a disciplined process to analyze the problem.

Problems in projects pass through three phases:

1. Define the problem.
2. Classify the problem.
3. Solve or neutralize the problem.

Defining the problem. Problem identification at first may seem trivial. But it becomes more involved when we ask what the problem really is, that is, what causes it. The first report we get of a problem will be a perception of the problem.* In simple problems, the problem definition is almost automatic with recognition of the problem. However in complex problems, the definition is more obscure. The difficulty lies in including our goal—what it is that we want to have happen—with the problem definition. Each person sees the problem differently. Usually, no one is completely wrong, nor is anyone totally right. Taking their perceptions as a whole provides a great amount of collective information. Therefore, perceptions are the logical place to start in defining a complex problem.

Whether the problem is simple or complex, we are looking for the cause of the problem as it is perceived. The cause is frequently pointed to by the different perceptions. Where it is not, it is likely to be more clearly defined at the next level down in the detail. The definition of a problem should be in terms of its cause, rather than in terms of a perception of it.

Classifying the problem. Appropriate solutions are heavily dependent on how we classify a problem. Many people see all problems alike. This perception is illustrated by the saying, "To a man with a hammer, all problems look like a nail." When we see all problems as simple, we have a tendency to guess at solutions. This behavior is attributable to an inability to cope with problems that are complicated. This also indicates an underlying inability to analyze problem changes comprehensively. Another flaw in problem solving is to use problem-solving meth-

---

* Nunn, P., *NSF Systems Approach for Management of Environmental Quality,* National Sanitation Foundation, Ann Arbor, Mich., 1973.

ods that have proved successful in the past, regardless of the type of problem. Whether these methods work will depend on how similar the current problem is to the one that is being used as a model.

Most of us one way or another need help with problem solving. We should not be limited in our ability to solve problems by oversimplification. For example, a common but complex problem is that of driving an automobile. Some people drive as if the solution to all problems is to apply the brake pedal. Fortunately, most drivers recognize the complexity of the decisions that need to be made, depending on the conditions that exist (unless we are mimicking what the person in front of us does). Driving on icy, slippery pavement requires different actions and maneuvers than those performed on dry pavement. Adjustments are necessary because at least one condition has changed. Complex problems are not uncommon, and we should come to recognize them.

There are several ways of classifying problems. Some are quite detailed and complex. But one simple method is shown in Fig. 9.7. One of the advantages of this four-cell matrix is that it helps us pick a problem-solving method which is appropriate to our problem. A simple problem-solving method applied to a complex problem will not yield a satisfying answer. Conversely, a complex problem-solving method applied to a simple problem is like killing a fly with a sledgehammer.

In Fig. 9.7, problems are classified by whether they are simple or complex and by whether they are malfunctions or improvements.

Simple or complex is probably the easier of the two pairs to understand. It refers to the type of system involved in the problem. Simple problems have a direct cause-and-effect relationship, similar to a broken wire causing a circuit failure. Complex problems have multiple interrelationships. This condition is illustrated by our previous example: applying the brakes on icy pavement does not give the same result as when the pavement is dry. The difference is not in the action of applying the brakes. It is in the condition of the road surface. The way the brakes are applied has to be changed to get a similar result. We don't just fix complex problems. Something has to be changed to get the result we want.

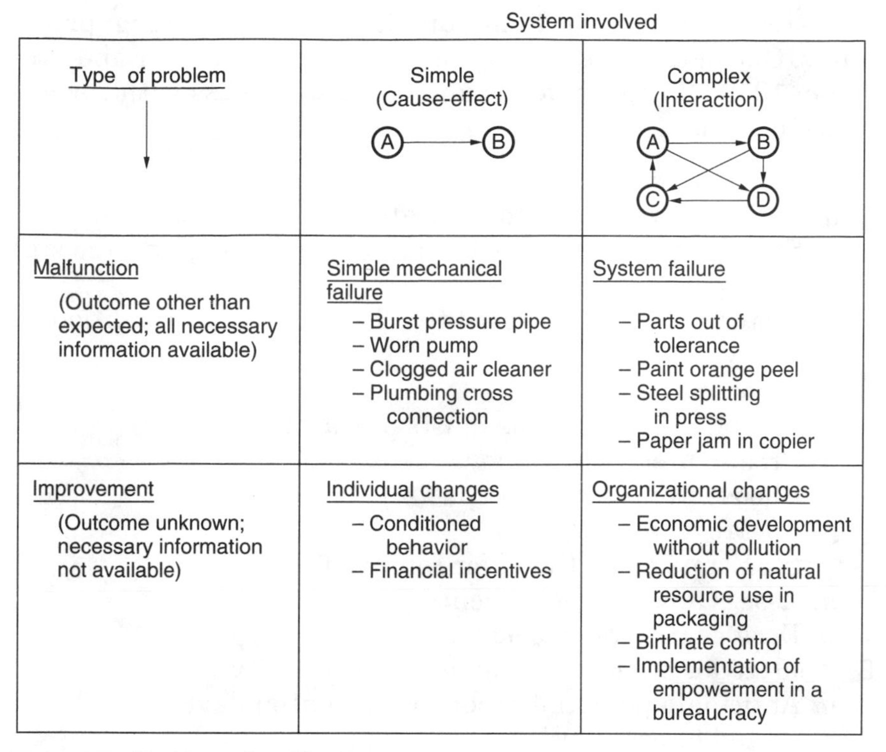

System involved

Type of problem ↓	Simple (Cause-effect)	Complex (Interaction)
Malfunction (Outcome other than expected; all necessary information available)	Simple mechanical failure – Burst pressure pipe – Worn pump – Clogged air cleaner – Plumbing cross connection	System failure – Parts out of tolerance – Paint orange peel – Steel splitting in press – Paper jam in copier
Improvement (Outcome unknown; necessary information not available)	Individual changes – Conditioned behavior – Financial incentives	Organizational changes – Economic development without pollution – Reduction of natural resource use in packaging – Birthrate control – Implementation of empowerment in a bureaucracy

Figure 9.7 Problem classification.

The malfunction or improvement pair is not so familiar to us, but it is critically important to our ability to solve problems efficiently and effectively. It refers to the type of problem. Malfunction problems are those where the system has been working correctly, then ceases to do so. For example, when your automobile fails to start one morning, you have a malfunction problem. You can try to solve it, or you can call in a mechanic to solve it. Either way, the principal characteristic of a malfunction problem is that all of the necessary information exists, even though it may not all be apparent to us. If we don't have the information on the malfunction which is needed to correct it, that is ignorance on our part—not a lacking in the system or the malfunction. We can dig it out, or it may be more apparent to someone more familiar with the system, such as a mechanic. Correcting a malfunction requires only that we return the system to its original operating condition.

There are some good malfunction problem-solving procedures. One of the better ones was developed by Kepner and Tregoe in 1958.* It recommends a four-phase process with several steps to each:

1. Problem analysis
   a. Recognize the problem (Deviation = Should – actual)
   b. Specify the deviation, what is working wrong, what is still working right
   c. Test for possible causes of the "is" and "is not" condition
   d. Verify the cause
2. Decision making
   a. Establish objectives as musts or wants
   b. Create alternative actions
   c. Choose from objectives
      (1) Musts are go/no-go
      (2) Wants are selected for relative fit
   d. Assess adverse consequences
   e. Decide on corrective action
3. Analyze for other potential problems
   a. Anticipate potential problems (Potential deviation = What should happen – What could happen)
   b. Anticipate possible causes of deviation in solution
4. Direction and control
   a. Take corrective action
   b. Set controls to detect recurrence
   c. Make contingency plans to minimize effects of recurrence

The other type of problem is the improvement problem. This is a different type of problem in that it is the situation in which we want to change the way a system is working. As a system, it is working just the way we expect it to, but we want to change that. The solution to an improvement problem requires innovation.

Unlike the malfunction problem where we are trying to correct a change which occurred in the system, in the improvement problem we are trying to make a controlled change in the system

---

* Kepner, C.H., and Tregoe, B.B., *The Rational Manager,* McGraw-Hill, New York, N.Y., 1965.

itself. To do this, we need information which we did not have previously. So, a learning process is implicit in the solution of improvement problems.

Almost all problems involving people are improvement problems because people are adaptable; that is, they adjust to changes. The simpler ones may involve an individual whose behavior we want to change. An example would be teaching someone how to read engineering drawings. The more complex problems involve making changes in how an organization does its work. An example would be institutionalizing the concept of total quality so it becomes a part of everything that's done.

There is no solution to a complex improvement problem. Problems that involve human behavior cannot be solved, but they can be controlled. The important challenge is how to modify the system so it operates in the desired way. However, we should recognize that complex human systems continue to change over time. We will have to make further modifications in the future. For example, when driving a car, there is no single action we perform which, when it ends, can be called "driving the car." Driving is comprised of a complex of continuing actions. If, at any time, we think we have finished and no longer exert control, we will surely stop driving.

Unlike the malfunction problem, all information is not available on an improvement problem. The answer to an improvement problem requires learning. Initially, we usually do not know enough to find a satisfactory action. The reason for this deficiency is that we seldom understand the system which is misbehaving well enough to make a change in it which eliminates our problem without creating another one which is just as bad.

Complex improvement problems are attacked with a broad-based systems approach which is capable of accounting for interactions. This implies dynamic modeling and simulation. There are several tools available, but one which is particularly well suited for project management problems is Systems Dynamics described by Jay Forrester in *Industrial Dynamics*.* Its application to projects can be found in *Introduction to Systems Dynam-*

---

* Forrester, J.W., *Industrial Dynamics,* MIT Press, Cambridge, Mass., 1961.

ics Modeling with Dynamo by Richardson and Pugh.* Systems Dynamics utilizes feedback loops and causal relationships to analyze the behavior of complex systems over time.

The difference between the solution to malfunction problems and improvement problems is that the malfunction solution returns us to the status quo, or previous condition, while the improvement solution brings us to a new situation.

Solving or neutralizing the problem. Another consideration in problem solving is the priority of the problem or its relative importance. All too frequently we assign top priority to a problem which is strongly argued by a team member or manager. Another weakness we have is that we will be attracted to problems which we know something about. Another type of problem solver is the crisis manager. These people are analogously referred to as fire fighters. My personal observation is that most industrial fire fighters are also arsonists. That is, they get their glory from fire fighting, so if there are no fires, they light one.

Interestingly, one of the simplest, yet effective, ways of prioritizing problems comes to us from the managers of large crises. Steven Fink gives us this method in his small but to-the-point book, *Crisis Management.*† Although his focus was on total business crises, this has been modified to fit project problems. The method is simply to place a problem on a four-quadrant matrix by rating its severity and its probability of occurring.

Since we are using probability of occurrence, we are performing this analysis before the problem occurs. Therefore, we must evaluate a problem as soon as we recognize its warning signs. In most projects, there are problems which can be evaluated in the preliminary planning. This activity falls in the risk management part of the project management body of knowledge.‡

The format of the problem priority analysis is shown in Fig. 9.8. The vertical axis is a rating of the expected severity of the problem. The horizontal axis is an estimate of the probability

---

* Richardson, G.P., and Pugh, A.L., Introduction to Systems Dynamics Modeling with Dynamo, MIT Press, Cambridge, Mass., 1983.

† Fink, Steven, Crisis Management, Planning for the Inevitable, Amacom, New York, N.Y., 1986.

‡ ———, *Project Management Body of Knowledge (PMBOK),* Project Management Institute, Drexel Hill, Pa., 1987.

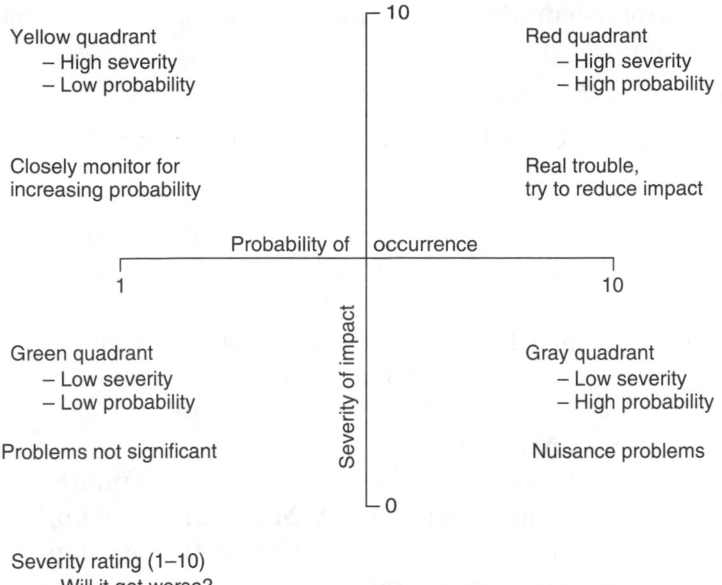

Figure 9.8 Problem priority matrix.

that the problem will occur. Problems occurring in the "green" quadrant have low severity and low probability, and need little attention. Problems occurring in the "yellow" quadrant have high severity but low probability, so need to be watched for signs of increasing probability. Conversely, problems in the "gray" quadrant have low severity but high probability, so they will occur but need to be endured as a nuisance. Problems which occur in the "red" quadrant have both high severity and high probability, so take your best shot—it may be the only one you get. In the "red" quadrant, your objective is to reduce the severity of the impending problem as much as you can. This action must be taken in one of the categories which make up the severity rating.

The severity rating is made up of five categories:

1. Will it get worse? Can it seriously damage or even kill the project?
2. Does it contain a health, safety, or legal hazard? Could someone get sick or injured? Is it illegal or unethical?

3. Does it compromise product performance? Does it degrade the quality of the product?
4. Is it going to cause a cost overrun?
5. Is it going to cause the scheduled completion to slip? Will the project be late?

The answers to these questions are judgmental. It is best if the team works on estimating the ratings. Each of the five questions is rated on a scale of 0 to 10. A rating of 0 represents no impact on the project. A rating of 10 represents as bad a condition as can be imagined. When the five severity categories have been rated, add the ratings together and divide by 5. Mark this score on the vertical "Severity of impact" scale.

After the severity of impact has been calculated, estimate the probability that the problem will occur. Mark this probability estimate on the horizontal axis. The position of the problem in the matrix is where these two scores intersect.

The quadrant in which the problem is plotted determines the problem's relative importance and its priority for attention. This simple technique will at least prevent the team from expending a lot of time and energy on trivial problems.

Evaluation

As the project progresses, we should make periodic checks on how well the process is working. We can close this evaluation loop on the project management process by rerunning the project implementation profile. It was first used to evaluate our baseline plan strength in Chap. 8. If we perform this evaluation periodically, such as quarterly, we can get an indication of whether there are any changes in the quality of the project management process. Changes such as strengthening of weaknesses are as important to us as categories which may have slipped over time and need more attention.

TQM Tracking

Our TQM implementation project has gotten off to a resounding start with the search for a consultant. We have a number of tools at our disposal to evaluate project schedule and status throughout its life cycle. To aid in the evaluation, we will use CA-SuperProject, an excellent software program by Computer Associates. In previous chapters we developed the project by hand in a fragmented fashion. The WBS, network diagram, resource analysis, etc., were prepared individually with no method of linking the information together. By using CA-SuperProject the component parts are automatically interconnected. Modifications in one area will be reflected in the others, making updates and what-if scenarios a simple exercise.

Figure 9.9 illustrates the information at the outset of the project, without linkages. Initially, all the tasks begin on January 1, 1993, and continue through their individual durations (lengths of the bars). After the finish-to-start links are input, the result is a Gantt chart as seen in Fig. 9.10. The solid bars represent the critical path which runs from phases 1 through 3; improve communication and hire date is April 27, 1993, well before the required December 31, 1993 management objective. It is easy to see the overall schedule of the project at a glance by using the Gantt chart.

Other useful information about the project is available for review. Figure 9.11 shows many of the items we discussed in previous chapters, some of which were calculated by hand. The work breakdown structure with the codes is created automatically as the tasks are input. Early and late starts and finishes are calculated based on task durations and linkages. The total float and free float are determined for each task, again via task duration and linkage calculations. A seeming myriad of information is at hand for use at any time during the project.

Figure 9.12 is the PERT chart, or logic network of the project. This information is basically the same as that found in the Gantt chart, but it is in a different form. Each of the task boxes contains an ID number, the scheduled duration, the task description, and the scheduled start and finish dates. The darker lines represent the critical path which corresponds to the dark bars on the Gantt chart.

We established a two-week reporting cycle for the project, and during the first two reporting periods the tasks went as scheduled. At the end of the third reporting period, on February 13, the consultant was hired, completing phase 3. At the end of the fourth reporting period, February 27, the following tasks were completed within their scheduled time frame:

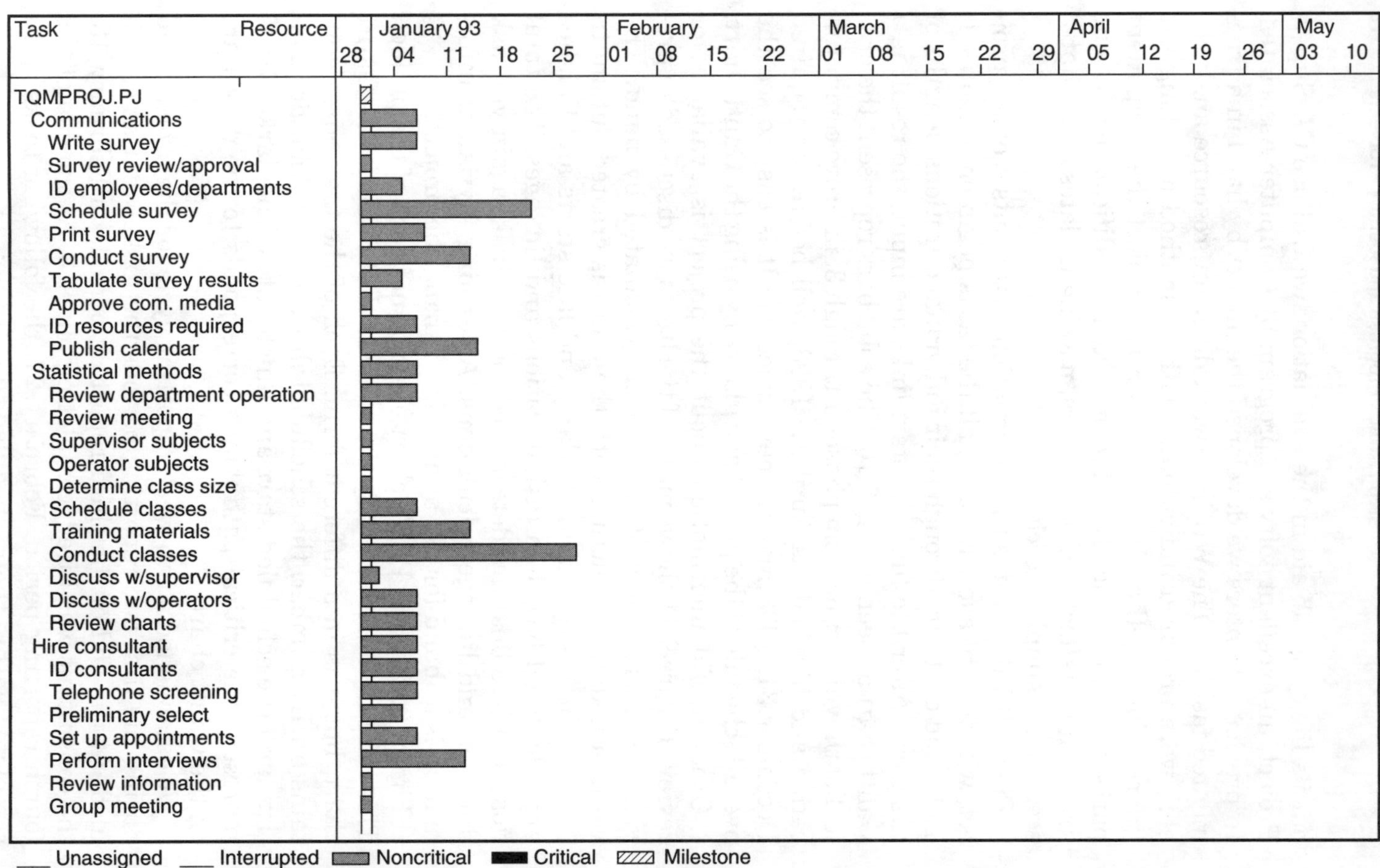

Figure 9.9 TQM task duration bar chart.

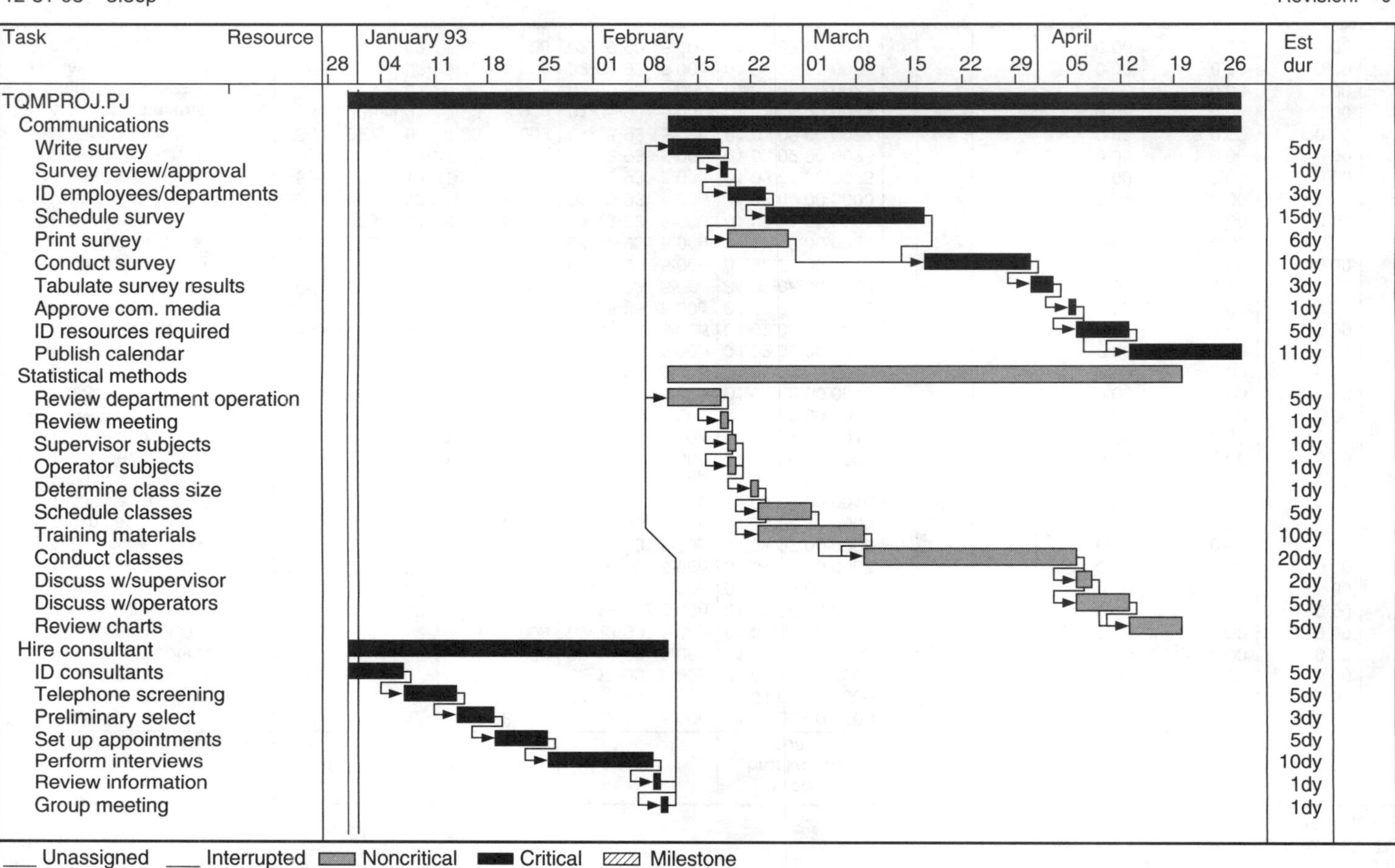

Figure 9.10 TQM Gantt chart.

GANTT-1
12-31-92 3:40p

Page 2 (2, 1)
Project: TQMPROJ.PJ
Revision: 0

Task	Resource	Sched dur	Task ID	Scheduled start	Scheduled finish	Work breakdown code	Pri	Alloc /day	Units assgn	Sched rsrc regular hrs	Sched rsrc overtime hrs	Sched rsrc conflict hrs
TQMPROJ.PJ		84dy	P1	12-31-92> 8:00a	04-27-93 5:00p	01.00.00.00.0000	50			0.00	0.00	0.00
Communications		54dy	001	02-11-93 8:00a	04-27-93 5:00p	01.01.00.00.0000				0.00	0.00	0.00
Write survey		5dy	002	02-11-93 8:00a	02-17-93 5:00p	01.01.01.00.0002				0.00	0.00	0.00
Survey review/approv		1dy	003	02-18-93 8:00a	02-18-93 5:00p	01.01.02.00.0003				0.00	0.00	0.00
ID employees/depts		3dy	004	02-19-93 8:00a	02-23-93 5:00p	01.01.03.00.0004				0.00	0.00	0.00
Schedule survey		15dy	005	02-24-93 8:00a	03-16-93 5:00p	01.01.04.00.0005				0.00	0.00	0.00
Print survey		6dy	006	02-19-93 8:00a	02-26-93 5:00p	01.01.05.00.0006				0.00	0.00	0.00
Conduct survey		10dy	007	03-17-93 8:00a	03-30-93 5:00p	01.01.06.00.0007				0.00	0.00	0.00
Tabulate survey reslt		3dy	008	03-31-93 8:00a	04-02-93 5:00p	01.01.07.00.0008				0.00	0.00	0.00
Approve com media		1dy	009	04-05-93 8:00a	04-05-93 5:00p	01.01.08.00.0009				0.00	0.00	0.00
ID resources req'd		5dy	010	04-06-93 8:00a	04-12-93 5:00p	01.01.09.00.0010				0.00	0.00	0.00
Publish calendar		11dy	011	04-13-93 8:00a	04-27-93 5:00p	01.01.10.00.0011				0.00	0.00	0.00
Statistical methods		48dy	012	02-11-93 8:00a	04-19-93 5:00p	01.02.00.00.0000				0.00	0.00	0.00
Review dept operation		5dy	013	02-11-93 8:00a	02-17-93 5:00p	01.02.01.00.0013				0.00	0.00	0.00
Review meeting		1dy	014	02-18-93 8:00a	02-18-93 5:00p	01.02.02.00.0014				0.00	0.00	0.00
Supervisor subjects		1dy	015	02-19-93 8:00a	02-19-93 5:00p	01.02.03.00.0015				0.00	0.00	0.00
Operator subjects		1dy	016	02-19-93 8:00a	02-19-93 5:00p	01.02.04.00.0016				0.00	0.00	0.00
Determine class size		1dy	017	02-22-93 8:00a	02-22-93 5:00p	01.02.05.00.0017				0.00	0.00	0.00
Schedule classes		5dy	018	02-23-93 8:00a	03-01-93 5:00p	01.02.06.00.0018				0.00	0.00	0.00
Training materials		10dy	019	02-23-93 8:00a	03-08-93 5:00p	01.02.07.00.0019				0.00	0.00	0.00
Conduct classes		20dy	020	03-09-93 8:00a	04-05-93 5:00p	01.02.08.00.0020				0.00	0.00	0.00
Discuss w/supervisor		2dy	021	04-06-93 8:00a	04-07-93 5:00p	01.02.09.00.0021				0.00	0.00	0.00
Discuss w/operators		5dy	022	04-06-93 8:00a	04-12-93 5:00p	01.02.10.00.0022				0.00	0.00	0.00
Review charts		5dy	023	04-13-93 8:00a	04-19-93 5:00p	01.02.11.00.0023				0.00	0.00	0.00
Hire consultant		30dy	024	12-31-93 8:00a	02-10-93 5:00p	01.03.00.00.0000				0.00	0.00	0.00
ID consultants		5dy	025	12-31-93 8:00a	01-06-93 5:00p	01.03.01.00.0025				0.00	0.00	0.00
Telephone screening		5dy	026	01-07-93 8:00a	01-13-93 5:00p	01.03.02.00.0026				0.00	0.00	0.00
Preliminary select		3dy	027	01-14-93 8:00a	01-18-93 5:00p	01.03.03.00.0027				0.00	0.00	0.00
Set up appointments		5dy	028	01-19-93 8:00a	01-25-93 5:00p	01.03.04.00.0028				0.00	0.00	0.00
Perform interviews		10dy	029	01-26-93 8:00a	02-08-93 5:00p	01.03.05.00.0029				0.00	0.00	0.00
Review information		1dy	030	02-09-93 8:00a	02-09-93 5:00p	01.03.06.00.0030				0.00	0.00	0.00
Group meeting		1dy	031	02-10-93 8:00a	02-10-93 5:00p	01.03.07.00.0031				0.00	0.00	0.00

Figure 9.11 TQM project details.

GANTT-1
12-31-92 3:41p

Page 3 (3, 1)
Project: TQMPROJ.PJ
Revision: 0

Task	Resource	Status	Task options	Task type	Account code	Early start	Early finish	Must start	Must finish
TQMPROJ.PJ						12-31-92 8:00a	04-27-93 5:00p	12-31-92> 8:00a	
Communications		Schd/crit		ASAP		02-11-93 8:00a	04-27-93 5:00p		
Write survey		Schd/crit	Resource	ASAP		02-11-93 8:00a	02-17-93 5:00p		
Survey review/approv		Schd/crit	Resource	ASAP		02-18-93 8:00a	02-18-93 5:00p		
ID employees/depts		Schd/crit	Resource	ASAP		02-19-93 8:00a	02-23-93 5:00p		
Schedule survey		Schd/crit	Resource	ASAP		02-24-93 8:00a	03-16-93 5:00p		
Print survey		Schd	Resource	ASAP		02-19-93 8:00a	02-26-93 5:00p		
Conduct survey		Schd/crit	Resource	ASAP		03-17-93 8:00a	03-30-93 5:00p		
Tabulate survey reslt		Schd/crit	Resource	ASAP		03-31-93 8:00a	04-02-93 5:00p		
Approve com media		Schd/crit	Resource	ASAP		04-05-93 8:00a	04-05-93 5:00p		
ID resources req'd		Schd/crit	Resource	ASAP		04-06-93 8:00a	04-12-93 5:00p		
Publish calendar		Schd/crit	Resource	ASAP		04-13-93 8:00a	04-27-93 5:00p		
Statistical methods		Schd		ASAP		02-11-93 8:00a	04-19-93 5:00p		
Review dept operation		Schd	Resource	ASAP		02-11-93 8:00a	02-17-93 5:00p		
Review meeting		Schd	Resource	ASAP		02-18-93 8:00a	02-18-93 5:00p		
Supervisor subjects		Schd	Resource	ASAP		02-19-93 8:00a	02-19-93 5:00p		
Operator subjects		Schd	Resource	ASAP		02-19-93 8:00a	02-19-93 5:00p		
Determine class size		Schd	Resource	ASAP		02-22-93 8:00a	02-22-93 5:00p		
Schedule classes		Schd	Resource	ASAP		02-23-93 8:00a	03-01-93 5:00p		
Training materials		Schd	Resource	ASAP		02-23-93 8:00a	03-08-93 5:00p		
Conduct classes		Schd	Resource	ASAP		03-09-93 8:00a	04-05-93 5:00p		
Discuss w/supervisor		Schd	Resource	ASAP		04-06-93 8:00a	04-07-93 5:00p		
Discuss w/operators		Schd	Resource	ASAP		04-06-93 8:00a	04-12-93 5:00p		
Review charts		Schd	Resource	ASAP		04-13-93 8:00a	04-19-93 5:00p		
Hire consultant		Schd/crit		ASAP		12-31-92 8:00a	02-10-93 5:00p		
ID consultants		Schd/crit	Resource	ASAP		12-31-92 8:00a	01-06-93 5:00p		
Telephone screening		Schd/crit	Resource	ASAP		01-07-93 8:00a	01-13-93 5:00p		
Preliminary select		Schd/crit	Resource	ASAP		01-14-93 8:00a	01-18-93 5:00p		
Set up appointments		Schd/crit	Resource	ASAP		01-19-93 8:00a	01-25-93 5:00p		
Perform interviews		Schd/crit	Resource	ASAP		01-26-93 8:00a	02-08-93 5:00p		
Review information		Schd/crit	Resource	ASAP		02-09-93 8:00a	02-09-93 5:00p		
Group meeting		Schd/crit	Resource	ASAP		02-10-93 8:00a	02-10-93 5:00p		

Figure 9.11 *(Continued)*

GANTT-1
12-31-92 3:42p

Page 4 (4, 1)
Project: TQMPROJ.PJ
Revision: 0

Task	Resource	Late start	Late finish	Start delay	Float	Free float	Baseline dur	Baseline start	Baseline finish	Basin rsrc total hrs	
TQMPROJ.PJ		12-31-92 8:00a	04-27-93 5:00p				0dy			0.00	
Communications		02-11-93 8:00a	04-27-93 5:00p	0dy	0dy	0dy	0dy			0.00	
Write survey		02-11-93 8:00a	02-17-93 5:00p	0dy	0dy	0dy	0dy			0.00	
Survey review/approv		02-18-93 8:00a	02-18-93 5:00p	0dy	0dy	0dy	0dy			0.00	
ID employees/depts		02-19-93 8:00a	02-23-93 5:00p	0dy	0dy	0dy	0dy			0.00	
Schedule survey		02-24-93 8:00a	03-16-93 5:00p	0dy	0dy	0dy	0dy			0.00	
Print survey		03-09-93 8:00a	02-16-93 5:00p	0dy	12dy	12dy	0dy			0.00	
Conduct survey		03-17-93 8:00a	03-30-93 5:00p	0dy	0dy	0dy	0dy			0.00	
Tabulate survey reslt		03-31-93 8:00a	04-02-93 5:00p	0dy	0dy	0dy	0dy			0.00	
Approve com media		04-05-93 8:00a	04-05-93 5:00p	0dy	0dy	0dy	0dy			0.00	
ID resources req'd		04-06-93 8:00a	04-12-93 5:00p	0dy	0dy	0dy	0dy			0.00	
Publish calendar		04-13-93 8:00a	04-27-93 5:00p	0dy	0dy	0dy	0dy			0.00	
Statistical methods		02-19-93 8:00a	04-27-93 5:00p	0dy	6dy	6dy	0dy			0.00	
Review dept operation		02-19-93 8:00a	02-25-93 5:00p	0dy	6dy	0dy	0dy			0.00	
Review meeting		02-26-93 8:00a	02-26-93 5:00p	0dy	6dy	0dy	0dy			0.00	
Supervisor subjects		03-01-93 8:00a	03-01-93 5:00p	0dy	6dy	0dy	0dy			0.00	
Operator subjects		03-01-93 8:00a	03-01-93 5:00p	0dy	6dy	0dy	0dy			0.00	
Determine class size		03-02-93 8:00a	03-02-93 5:00p	0dy	6dy	0dy	0dy			0.00	
Schedule classes		03-10-93 8:00a	03-16-93 5:00p	0dy	11dy	5dy	0dy			0.00	
Training materials		03-03-93 8:00a	03-16-93 5:00p	0dy	6dy	0dy	0dy			0.00	
Conduct classes		03-17-93 8:00a	04-13-93 5:00p	0dy	6dy	0dy	0dy			0.00	
Discuss w/supervisor		04-19-93 8:00a	04-20-93 5:00p	0dy	9dy	3dy	0dy			0.00	
Discuss w/operators		04-14-93 8:00a	04-20-93 5:00p	0dy	6dy	0dy	0dy			0.00	
Review charts		04-21-93 8:00a	04-27-93 5:00p	0dy	6dy	6dy	0dy			0.00	
Hire consultant		12-31-92 8:00a	02-10-93 5:00p	0dy	0dy	0dy	0dy			0.00	
ID consultants		12-31-92 8:00a	01-06-93 5:00p	0dy	0dy	0dy	0dy			0.00	
Telephone screening		01-07-93 8:00a	01-13-93 5:00p	0dy	0dy	0dy	0dy			0.00	
Preliminary select		01-14-93 8:00a	01-18-93 5:00p	0dy	0dy	0dy	0dy			0.00	
Set up appointments		01-19-93 8:00a	01-25-93 5:00p	0dy	0dy	0dy	0dy			0.00	
Perform interviews		01-26-93 8:00a	02-08-93 5:00p	0dy	0dy	0dy	0dy			0.00	
Review information		02-09-93 8:00a	02-09-93 5:00p	0dy	0dy	0dy	0dy			0.00	
Group meeting		02-10-93 8:00a	02-10-93 5:00p	0dy	0dy	0dy	0dy			0.00	

Figure 9.11 *(Continued)*

GANTT-1
12-31-92 3:43p

Page 5 (5, 1)
Project: TQMPROJ.PJ
Revision: 0

Task	Resource	Description	
TQMPROJ.PJ			
Communications			
Write survey			
Survey review/approv			
ID employees/depts			
Schedule survey			
Print survey			
Conduct survey			
Tabulate survey reslt			
Approve com media			
ID resources req'd			
Publish calendar			
Statistical methods			
Review dept operation			
Review meeting			
Supervisor subjects			
Operator subjects			
Determine class size			
Schedule classes			
Training materials			
Conduct classes			
Discuss w/supervisor			
Discuss w/operators			
Review charts			
Hire consultant			
ID consultants			
Telephone screening			
Preliminary select			
Set up appointments			
Perform interviews			
Review information			
Group meeting			

Figure 9.11 *(Continued)*

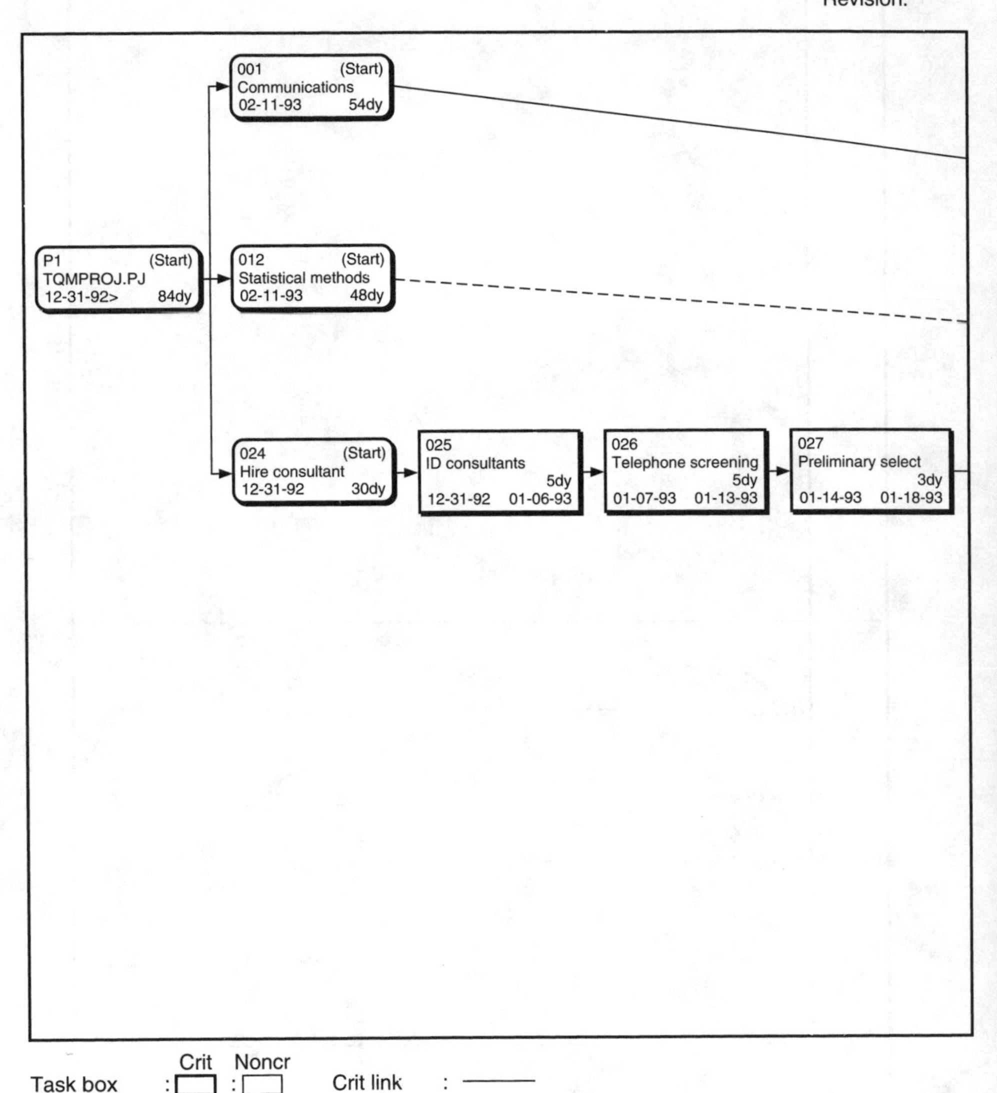

Figure 9.12 TQM PERT chart.

Pert 3
12-31-92 3:14p

Page 2 (2, 1)
Project: TQMPROJ.PJ
Revision:

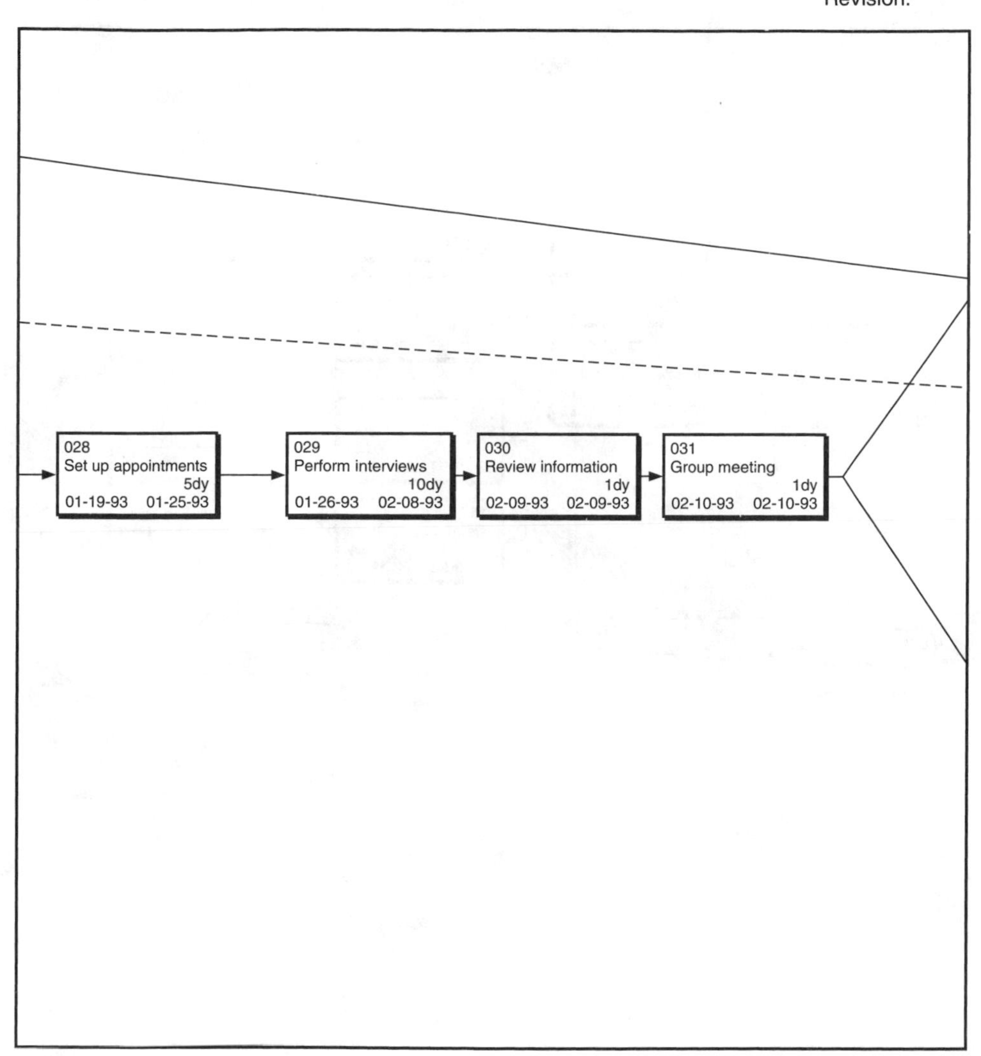

	Crit	Noncr
Task box	: ▭	: ▭
Heading box	: ⬭	: ⬭
Milestone box	: ▭	: ▭

Crit link : ———
Noncrit link : - - - - -

Figure 9.12 *(Continued)*

Pert 3 Page 3 (3, 1)
12-31-92 3:18p Project: TQMPROJ.PJ
Revision:

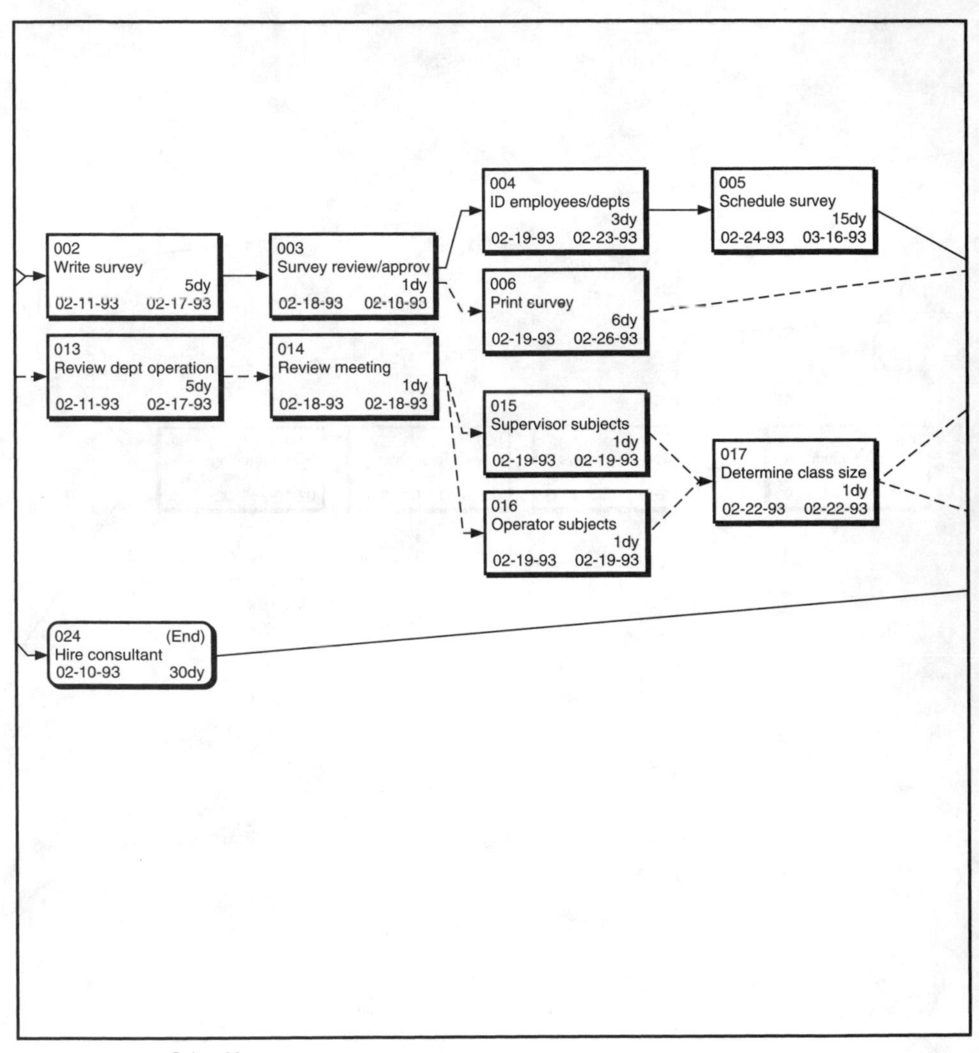

	Crit	Noncr
Task box	: ▭	: ▭
Heading box	: ⬭	: ⬭
Milestone box	: ▭	: ▭

Crit link : ———
Noncrit link : - - - - -

Figure 9.12 *(Continued)*

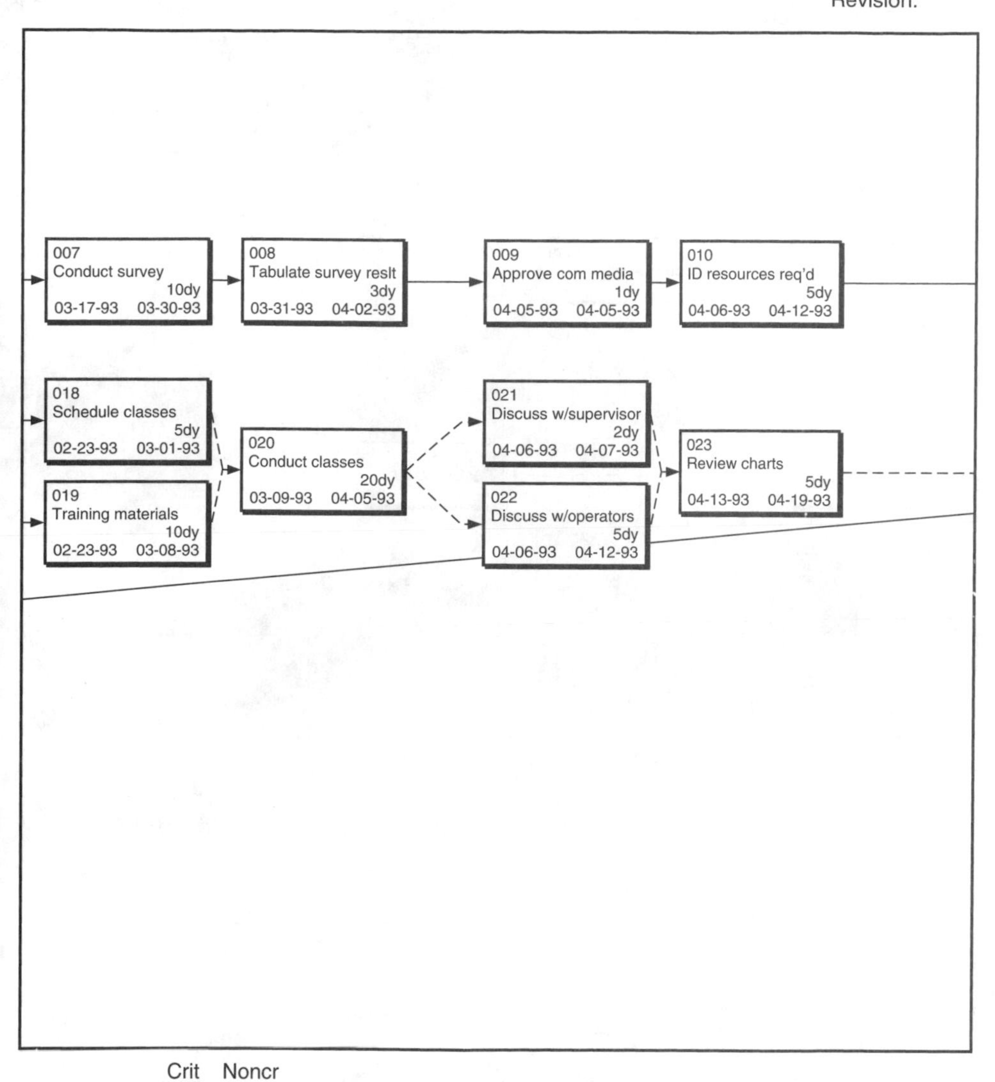

Figure 9.12 *(Continued)*

Pert 3
12-31-92 3:22p

Page 5 (5, 1)
Project: TQMPROJ.PJ
Revision:

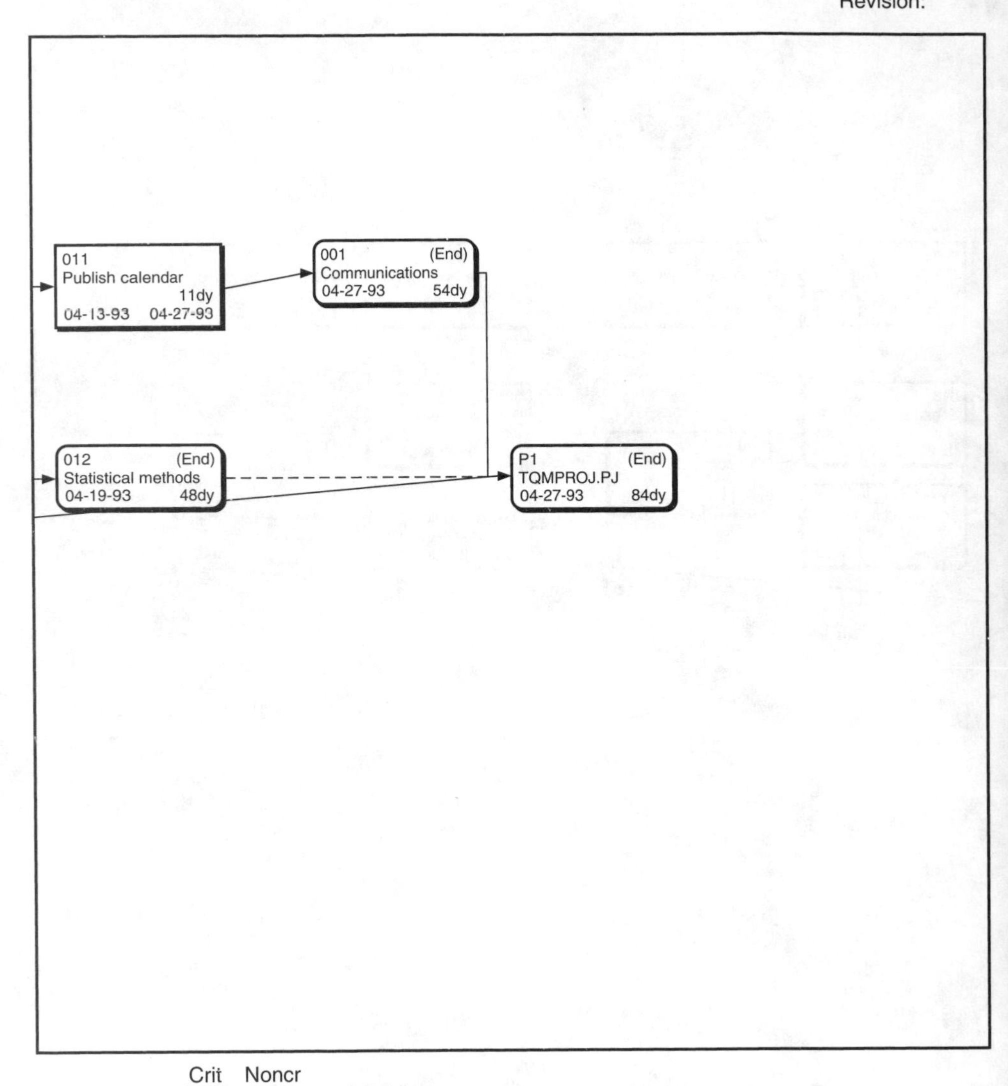

	Crit	Noncr
Task box	: ▭	: ▭
Heading box	: ⬭	: ⬭
Milestone box	: ▭	: ▭

Crit link : ———
Noncrit link : - - - - -

Figure 9.12 *(Continued)*

Status	*WBS code**	*Task*
OK	01.01.01	Write survey
OK	01.02.01	Review dept. operation
OK	01.02.02	Review meeting
OK	01.01.02	Survey review/approval
OK	01.02.03	Supervisor subjects
OK	01.02.04	Operator subjects
OK	01.02.05	Class size
OK	01.01.03	ID employees

* The WBS code used here is the one generated by the computer, not the one from previous chapters.

March 13 marks the end of fifth reporting period, but this time a problem has surfaced.

Status	*WBS code**	*Task*
OK	01.01.05	Print survey
OK	01.02.06	Schedule classes
Extended	01.02.07	Training materials

* The WBS code used here is the one generated by the computer, not the one from previous chapters.

There was a mechanical breakdown at the print shop which will cause a 10-day delay of this task. We need to immediately determine the impact of the problem on the project. At first glance, it appears as though the impact will be minor, as this task is not on the critical path. When we reenter the task duration as 20 days, however, both the total float and free float numbers become negative (see Fig. 9.13). Further analysis of the situation reveals that our critical path has changed and the total duration for the project has been lengthened (see Figs. 9.14 and 9.15). The project owner, C. P. Kaye, has some important decisions to make regarding the situation.

The project leader (the QA manager in this case) informs the project owner, C. P. Kaye, about the dilemma. Together they decide to meet with the project team to work on a solution to the problem.

First, the problem needs to be defined. The training materials which were to be used in the SPC classes cannot be produced at the scheduled time. In more detail, the printer experienced a

Training materials		Start	Finish
Duration: 20dy	Schedule Dur: 20dy	Schd: 02-23-93 8:00a	03-22-93 5:00p
Options: Resource	Type: ASAP	Must:	
% complete: 0 Pri:	Actual dur: 0dy	Actl:	
Delay: 0dy Free float: –10dy	Float: –4dy	Erly: 02-23-93 8:00a	03-08-93 5:00p
SD: 0dy 0pt: 20dy Like: 20dy	Pess: 20dy	Late: 03-03-93 8:00a	03-16-93 5:00p
WBS: 01.02.07.00.0019 schd		Base:	

Resource name	Schd dur	Scheduled start	Scheduled finish	Pri	Alloc /day	Units assign	Sched rsrc total hrs	Sched rsrc regular hrs	Sched rsrc overtime hrs	Sched rsrc conflict hrs	Account code	Early start	Early finish
Buyer/CS	10dy	02-23-93 8:00a	03-08-93 5:00p	50	100%	1	160.00	160.00	0.00	0.00	01.00.00.00.0000	02-23-93 8:00a	03-08-93 5:00p
Total							160.00	160.00	0.00	0.00			

Figure 9.13 Training materials task duration change.

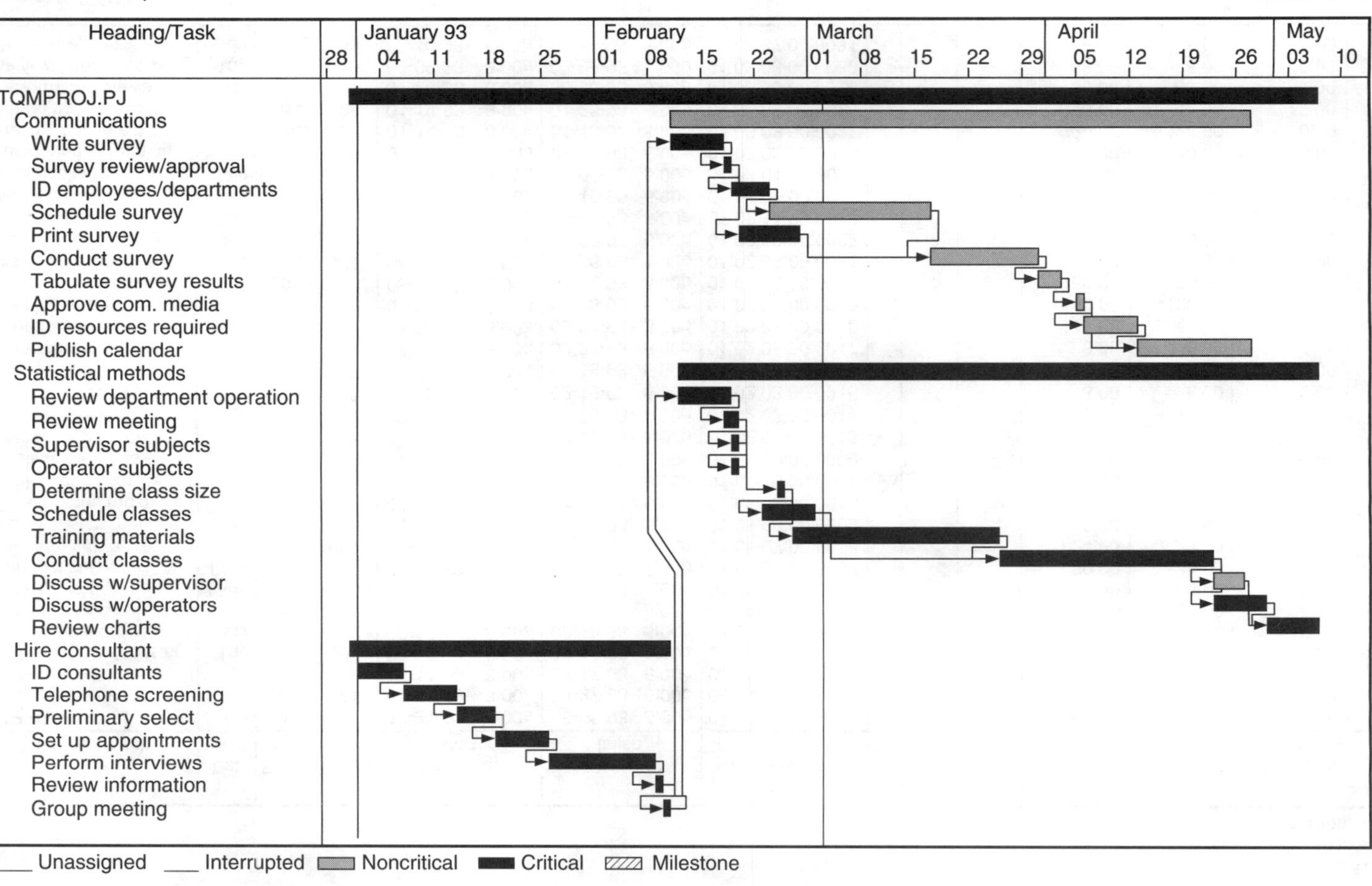

Figure 9.14 TQM Gantt chart revision.

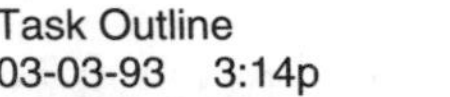

Task Outline
03-03-93 3:18p

Page 2 (2, 1)
Project: TQMPROJ.PJ
Revision: 1

Heading/task	Est dur	Sched dur	Task ID	Scheduled start	Scheduled finish	Work breakdown code	Pri	A!loc /day	Units assgn	Sched rsrc total hrs	Sched rsrc regular hrs	Sched rsrc overtime hrs
TQMPROJ.PJ		91dy	P1	12-31-92> 8:00a	05-06-93 5:00p	01.00.00.00.0000	50			1248.00	1248.00	0.00
Communications		54dy	001	02-11-93 8:00a	04-27-93 5:00p	01.01.00.00.0000				472.00	472.00	0.00
Write survey	5dy	5dy	002	02-11-93 8:00a	02-17-93 5:00p	01.01.01.00.0002				40.00	40.00	0.00
Survey review/approv	1dy	1dy	003	02-18-93 8:00a	02-18-93 5:00p	01.01.02.00.0003				8.00	8.00	0.00
ID employees/depts	3dy	3dy	004	02-19-93 8:00a	02-23-93 5:00p	01.01.03.00.0004				24.00	24.00	0.00
Schedule survey	15dy	15dy	005	02-24-93 8:00a	03-16-93 5:00p	01.01.04.00.0005				120.00	120.00	0.00
Print survey	6dy	5dy	006	02-20-93 8:00a	02-27-93 8:00a	01.01.05.00.0006				40.00	40.00	0.00
Conduct survey	10dy	10dy	007	03-17-93 8:00a	03-30-93 5:00p	01.01.06.00.0007				80.00	80.00	0.00
Tabulate survey reslt	3dy	3dy	008	03-31-93 8:00a	04-02-93 5:00p	01.01.07.00.0008				24.00	24.00	0.00
Approve com media	1dy	1dy	009	04-05-93 8:00a	04-05-93 5:00p	01.01.08.00.0009				8.00	8.00	0.00
ID resources req'd	5dy	5dy	010	04-06-93 8:00a	04-12-93 5:00p	01.01.09.00.0010				40.00	40.00	0.00
Publish calendar	11dy	11dy	011	04-13-93 8:00a	04-27-93 5:00p	01.01.10.00.0011				88.00	88.00	0.00
Statistical methods		60dy	012	02-12-93 8:00a	05-06-93 5:00p	01.02.00.00.0000				536.00	536.00	0.00
Review dept operation	5dy	5dy	013	02-12-93 8:00a	02-18-93 5:00p	01.02.01.00.0013				40.00	40.00	0.00
Review meeting	1dy	2dy	014	02-18-93 8:00a	02-19-93 5:00p	01.02.02.00.0014				16.00	16.00	0.00
Supervisor subjects	1dy	1dy	015	02-19-93 8:00a	02-19-93 5:00p	01.02.03.00.0015				8.00	8.00	0.00
Operator subjects	1dy	1dy	016	02-19-93 8:00a	02-19-93 5:00p	01.02.04.00.0016				8.00	8.00	0.00
Determine class size	1dy	1dy	017	02-25-93 8:00a	02-25-93 5:00p	01.02.05.00.0017				8.00	8.00	0.00
Schedule classes	5dy	5dy	018	02-23-93 8:00a	03-01-93 5:00p	01.02.06.00.0018				40.00	40.00	0.00
Training materials	20dy	20dy	019	02-26-93 8:00a	03-25-93 5:00p	01.02.07.00.0019				160.00	160.00	0.00
Conduct classes	20dy	20dy	020	03-26-93 8:00a	04-22-93 5:00p	01.02.08.00.0020				160.00	160.00	0.00
Discuss w/supervisor	2dy	2dy	021	04-23-93 8:00a	04-26-93 5:00p	01.02.09.00.0021				16.00	16.00	0.00
Discuss w/operators	5dy	5dy	022	04-23-93 8:00a	04-29-93 5:00p	01.02.10.00.0022				40.00	40.00	0.00
Review charts	5dy	5dy	023	04-30-93 8:00a	05-06-93 5:00p	01.02.11.00.0023				40.00	40.00	0.00
Hire consultant		30dy	024	12-31-93 8:00a	02-10-93 5:00p	01.03.00.00.0000				240.00	240.00	0.00
ID consultants	5dy	4dy	025	01-01-93 8:00a	01-06-93 5:00p	01.03.01.00.0025				40.00	40.00	0.00
Telephone screening	5dy	5dy	026	01-07-93 8:00a	01-13-93 5:00p	01.03.02.00.0026				40.00	40.00	0.00
Preliminary select	3dy	3dy	027	01-14-93 8:00a	01-18-93 5:00p	01.03.03.00.0027				24.00	24.00	0.00
Set up appointments	5dy	5dy	028	01-19-93 8:00a	01-25-93 5:00p	01.03.04.00.0028				40.00	40.00	0.00
Perform interviews	10dy	10dy	029	01-26-93 8:00a	02-08-93 5:00p	01.03.05.00.0029				80.00	80.00	0.00
Review information	1dy	1dy	030	02-09-93 8:00a	02-09-93 5:00p	01.03.06.00.0030				8.00	8.00	0.00
Group meeting	1dy	1dy	031	02-10-93 8:00a	02-10-93 5:00p	01.03.07.00.0031				8.00	8.00	0.00

Figure 9.15 TQM project details revision.

Task Outline
03-03-93 3:20p

Page 3 (3, 1)
Project: TQMPROJ.PJ
Revision: 1

Heading/task	Sched rsrc conflict hrs	Status	Task options	Task type	Account code	Early start	Early finish	Must start	
TQMPROJ.PJ	0.00					12-31-92 8:00a	05-06-93 5:00p	12-31-92> 8:00a	
Communications	0.00	Prog		ASAP		02-11-93 8:00a	04-27-93 5:00p		
Write survey	0.00	Done/crit	Resource	ASAP		02-11-93 8:00a	02-17-93 5:00p		
Survey review/approv	0.00	Done/crit	Resource	ASAP		02-18-93 8:00a	02-18-93 5:00p		
ID employees/depts	0.00	Done/crit	Resource	ASAP		02-19-93 8:00a	02-23-93 5:00p		
Schedule survey	0.00	Prog	Resource	ASAP		02-24-93 8:00a	03-16-93 5:00p		
Print survey	0.00	Done/crit	Resource	ASAP		02-19-93 8:00a	02-27-93 8:00a		
Conduct survey	0.00	Schd	Resource	ASAP		03-17-93 8:00a	03-30-93 5:00p		
Tabulate survey reslt	0.00	Schd	Resource	ASAP		03-31-93 8:00a	04-02-93 5:00p		
Approve com media	0.00	Schd	Resource	ASAP		04-05-93 8:00a	04-05-93 5:00p		
ID resources req'd	0.00	Schd	Resource	ASAP		04-06-93 8:00a	04-12-93 5:00p		
Publish calendar	0.00	Schd	Resource	ASAP		04-13-93 8:00a	04-27-93 5:00p		
Statistical methods	0.00	Prog/crit		ASAP		02-11-93 8:00a	05-06-93 5:00p		
Review dept operation	0.00	Done/crit	Resource	ASAP		02-11-93 8:00a	02-18-93 5:00p		
Review meeting	0.00	Done/crit	Resource	ASAP		02-18-93 8:00a	02-19-93 5:00p		
Supervisor subjects	0.00	Done/crit	Resource	ASAP		02-19-93 8:00a	02-19-93 5:00p		
Operator subjects	0.00	Done/crit	Resource	ASAP		02-19-93 8:00a	02-19-93 5:00p		
Determine class size	0.00	Done/crit	Resource	ASAP		02-22-93 8:00a	02-25-93 5:00p		
Schedule classes	0.00	Done/crit	Resource	ASAP		02-23-93 8:00a	03-01-93 5:00p		
Training materials	0.00	Schd/crit	Resource	ASAP		02-26-93 8:00a	03-25-93 5:00p		
Conduct classes	0.00	Schd/crit	Resource	ASAP		03-26-93 8:00a	04-22-93 5:00p		
Discuss w/supervisor	0.00	Schd	Resource	ASAP		04-23-93 8:00a	04-26-93 5:00p		
Discuss w/operators	0.00	Schd/crit	Resource	ASAP		04-23-93 8:00a	04-29-93 5:00p		
Review charts	0.00	Schd/crit	Resource	ASAP		04-30-93 8:00a	05-06-93 5:00p		
Hire consultant	0.00	Done/crit		ASAP		12-31-92 8:00a	02-10-93 5:00p		
ID consultants	0.00	Done/crit	Resource	ASAP		12-31-92 8:00a	01-06-93 5:00p		
Telephone screening	0.00	Done/crit	Resource	ASAP		01-07-93 8:00a	01-13-93 5:00p		
Preliminary select	0.00	Done/crit	Resource	ASAP		01-14-93 8:00a	01-18-93 5:00p		
Set up appointments	0.00	Done/crit	Resource	ASAP		01-19-93 8:00a	01-25-93 5:00p		
Perform interviews	0.00	Done/crit	Resource	ASAP		01-26-93 8:00a	02-08-93 5:00p		
Review information	0.00	Done/crit	Resource	ASAP		02-09-93 8:00a	02-09-93 5:00p		
Group meeting	0.00	Done/crit	Resource	ASAP		02-10-93 8:00a	02-10-93 5:00p		

Figure 9.15 (*Continued*)

Task Outline
03-03-93 3:21p

Page 4 (4, 1)
Project: TQMPROJ.PJ
Revision: 1

Heading/task	Must finish	Late start	Late finish	Start delay	Float	Free float	Baseln dur	Baseline start	Baseline finish	Basin rsrc total hrs
TQMPROJ.PJ		12-31-92 8:00a	05-06-93 5:00p				0dy			0.00
Communications		02-11-93 8:00a	05-06-93 5:00p	0dy	7dy	7dy	0dy			0.00
Write survey		02-11-93 8:00a	02-17-93 5:00p	0dy	0dy	0dy	0dy			0.00
Survey review/approv		02-18-93 8:00a	02-18-93 5:00p	0dy	0dy	0dy	0dy			0.00
ID employees/depts		02-19-93 8:00a	02-23-93 5:00p	0dy	0dy	0dy	0dy			0.00
Schedule survey		02-24-93 8:00a	03-25-93 5:00p	0dy	7dy	0dy	0dy			0.00
Print survey		02-20-93 8:00a	02-27-93 8:00a	0dy	0dy	0dy	0dy			0.00
Conduct survey		03-26-93 8:00a	04-08-93 5:00p	0dy	7dy	0dy	0dy			0.00
Tabulate survey reslt		04-09-93 8:00a	04-13-93 5:00p	0dy	7dy	0dy	0dy			0.00
Approve com media		04-14-93 8:00a	04-14-93 5:00p	0dy	7dy	0dy	0dy			0.00
ID resources req'd		04-15-93 8:00a	04-21-93 5:00p	0dy	7dy	0dy	0dy			0.00
Publish calendar		04-22-93 8:00a	05-06-93 5:00p	0dy	7dy	7dy	0dy			0.00
Statistical methods		02-12-93 8:00a	05-06-93 5:00p	0dy	0dy	0dy	0dy			0.00
Review dept operation		02-12-93 8:00a	02-18-93 5:00p	0dy	0dy	–1dy	0dy			0.00
Review meeting		02-18-93 8:00a	02-19-93 5:00p	0dy	0dy	–1dy	0dy			0.00
Supervisor subjects		02-19-93 8:00a	02-19-93 5:00p	0dy	0dy	0dy	0dy			0.00
Operator subjects		02-19-93 8:00a	02-19-93 5:00p	0dy	0dy	0dy	0dy			0.00
Determine class size		02-25-93 8:00a	02-25-93 5:00p	0dy	0dy	–3dy	0dy			0.00
Schedule classes		02-23-93 8:00a	03-01-93 5:00p	0dy	0dy	0dy	0dy			0.00
Training materials		02-26-93 8:00a	03-25-93 5:00p	0dy	0dy	0dy	0dy			0.00
Conduct classes		03-26-93 8:00a	04-22-93 5:00p	0dy	0dy	0dy	0dy			0.00
Discuss w/supervisor		04-28-93 8:00a	04-29-93 5:00p	0dy	3dy	3dy	0dy			0.00
Discuss w/operators		04-23-93 8:00a	04-29-93 5:00p	0dy	0dy	0dy	0dy			0.00
Review charts		04-30-93 8:00a	05-06-93 5:00p	0dy	0dy	0dy	0dy			0.00
Hire consultant		12-31-92 8:00a	02-10-93 5:00p	0dy	0dy	0dy	0dy			0.00
ID consultants		12-31-92 8:00a	01-06-93 5:00p	0dy	0dy	0dy	0dy			0.00
Telephone screening		01-07-93 8:00a	01-13-93 5:00p	0dy	0dy	0dy	0dy			0.00
Preliminary select		01-14-93 8:00a	01-18-93 5:00p	0dy	0dy	0dy	0dy			0.00
Set up appointments		01-19-93 8:00a	01-25-93 5:00p	0dy	0dy	0dy	0dy			0.00
Perform interviews		01-26-93 8:00a	02-08-93 5:00p	0dy	0dy	0dy	0dy			0.00
Review information		02-09-93 8:00a	02-09-93 5:00p	0dy	0dy	0dy	0dy			0.00
Group meeting		02-10-93 8:00a	02-10-93 5:00p	0dy	0dy	0dy	0dy			0.00

Figure 9.15 *(Continued)*

Task Ouline
06-22-93 4:14p

Page 5 (5, 1)
Project: TQMPROJ.PJ
Revision: 2

Heading/task	Description	
TQMPROJ.PJ		
Communications		
Write survey		
Survey review/approv		
ID employees/depts		
Schedule survey		
Print survey		
Conduct survey		
Tabulate survey reslt		
Approve com media		
ID resources req'd		
Publish calendar		
Statistical methods		
Review dept operation		
Review meeting		
Supervisor subjects		
Operator subjects		
Determine class size		
Schedule classes		
Training materials	New task duration—mechanical problems @ printer	
Conduct classes		
Discuss w/supervisor		
Discuss w/operators		
Review charts		
Hire consultant		
ID consultants		
Telephone screening		
Preliminary select		
Set up appointments		
Perform interviews		
Review information		
Group meeting		

Figure 9.15 *(Continued)*

mechanical breakdown at his production location which moved his schedule back.

The problem needs to be classified in terms of simple/complex and malfunction/creative. The system was working properly up to a point, and then it ceased to function. This scenario would make the problem a malfunction. Although we do not know the details of the damage, it is fair to assume that this was a simple mechanical failure. The interesting part of the problem is its actual location with respect to the project. By location we do not mean the address of the printer. The problem is connected to the project but located in the external subsystem of the business system (see Chap. 3, Fig. 3.4). We do not know if a more careful selection of the supplier would have prevented this problem. It does, however, reemphasize the necessity of scrutinizing external as well as internal project components.

The project team needs to evaluate the severity of the problem as well. The five severity questions are posed and then answered.

Q: Can it seriously damage or kill the project?
Will it get worse?

A: If left unattended it could get worse. (7 points)

Q: Does it contain a health, safety, or legal hazard?

A: It does not create any of these situations for the company implementing TQM. (0 points)

Q: Does it compromise product performance?

A: No. (0 points)

Q: Is it going to cause a cost overrun?

A: There is the potential for a cost overrun, depending on the final solution. (8 points)

Q: Will the project be late?

A: If the new schedule estimate is correct, yes. (10 points)

Average = 5.0 points

Refer to Fig. 9.8, the problem priority analysis we need to estimate the probability of occurrence. The problem has already taken place, so we will evaluate the probability based on the affected severity categories. What is the probability that there will be a cost overrun? What is the probability the project will be late? There would appear to be a moderately high probability of project delay and cost overrun. Giving the probability rating a 7, the entire analysis is very close to the "red" quadrant. Some type of action needs to be taken by the project team.

After deliberating for several hours, the project team came up with a number of potential solutions to the problem. By the way, the buyer was 20 minutes late for the meeting and was required to deposit $10.00 in the tardy pool.

1. Status quo; no action will be taken and the schedule will slip.
2. Inform top management of the problem and let them decide what to do.
3. Search for an alternative printer to do the task.
4. Try to do the printing in-house via copying machines.

The schedule will slip only a small amount of time, but there may be other projects queued after the finish of this one. Top management may or may not be receptive to the circumstances, but other options should be investigated initially. Trying to do the printing in-house may create more confusion with new resources, even if they are available. The project team decided to immediately look for another printer. Only when the search is completed will we know if the schedule will really slip. The cost will certainly increase due to the search activities, but there may or may not be a premium paid for the materials with a new printer.

Chapter

10

The Transition to Use

Moving the Project into Practice

The last step in integrating a program into the management system of a company is the transition from development and implementation to acceptance and use by the organization. This is the point at which the development team surrenders their work to the users. If the previous steps were performed with user participation, this final transition is only a formality.

At the end of development and implementation there are some items which must be accounted for, such as fulfillment of the objectives, a good communications interface between the development team and the users, closing of vendor contracts, recording the lessons learned, thanking the development team, and turning off the money. Each of these activities is important to the long-term success of this project and to the efficiency of subsequent projects.

Fulfillment of objectives

As stated for step 1 in Chap. 4, we know our project is done when we have fulfilled its objectives. In metaphorical terms, the death warrant for a project is written at its birth. When all the objectives laid out in the first step have been fulfilled, the project is completed.

At completion of a development or implementation project, there should be a formal review of the objectives and how they were accomplished. This review should include executive management and the users. When they are satisfied that we have accomplished our objectives, we are done.

Handing off to users

The development team's work is complete. Their deliverable must be given to the people who will be using it. This means giving them control and providing close support for a reasonable length of time. No matter how thorough the development, there will always be a need for some change to satisfy the users. This is especially true for the development of new policies like TQM.

The implementation of new policies, programs, and processes must be followed by development support for at least one operating cycle. There will be a need for changes as implementation is started throughout the company. These changes must be made so they are consistent with the intent of the new policy. This is especially important in the introduction of new policies because there will be a strong tendency for some managers to cling to the old way.

Closing vendor contracts

This is a purchasing function, but it requires that vendors be notified that the project is complete. Any open issues must be negotiated to a close. Good vendor performance should be acknowledged and perhaps rewarded.

Recording the lessons learned

Lessons learned include how the program was implemented as well as technical documentation. Documentation of a completed program is as important as the documentation of a completed product design. It will be used by subsequent projects as the template for their program. The plan for the new program will be developed from a markup of the current program plan. If, during this markup process, there is an intent to improve on the schedule, cost, quality, or to reduce risk, then it will be continuous improvement at work. Continuous improvement is achieved by improving on the previous program. Targets can be established for improvement.

An important form of improvement is to avoid the major problems of a previous program. Major problems seldom have simple causes, so it is important to document the causes and solutions for each major project problem. Using this information, subsequent programs may be able to plan their avoidance of the problem.

A system must be established to facilitate the passing on of lessons learned from one project to another. This continuity is needed to avoid reinventing the wheel and making the same mistakes in each new project.

Thanking the project team

Thanking the project team may seem like a trivial act; however, it is one of the most important. Projects are not routine work because they are intended to accomplish something that has not been done before. They frequently require people to stretch their abilities in order to solve problems, or to make up time lost solving unanticipated problems. Project work is not easy. It takes a lot of hard work and some extra effort. Acknowledging the contribution of each person is important to that person and to the level of his or her future performance for the company.

A human aspect of project work which is often overlooked is the depressing of morale caused by apprehension at the end of a project. When people don't know where they are going next, they tend to get apprehensive.* This generally causes morale to drop. When morale drops, productivity suffers. The solution to this problem is to find an assignment for each of the people on the team prior to the end of the project and let them know about it.

Turning off the money

The old admonition is for the last person out to turn off the lights. Similarly, the last act in closing down a project is to turn off the money. It's more than a symbolic gesture. It stops all

---

* Stork, D., and Sapienza, A., "Task and Human Messages Over the Project Life Cycle: Matching Media to Messages," *Project Management Journal,* Project Management Institute, Drexel Hill, Pa., December 1992.

work. This is necessary because there are occasionally people spending project funds clandestinely. This act ends it all.

At one time, I was asked by a company financial manager how to determine which of his 565 projects were real. He suspected that about half of them were not valid projects, but couldn't separate them. My suggestion was to first separate the alive ones from the dead. But it would require him to make some apologies for inconveniences. The plan was to turn off the money for all projects. The managers of those which were active would complain quickly. He would have to apologize to them and restore their money. Some not-so-active projects might take a week to discover their loss of funding. They too could have their money restored after they complained. A third class of project managers would never complain because their projects were not valid anyway. These projects would be killed.

He did it. In two weeks' time, he had reduced his project load from 565 to 160. The last time I heard from him, his project load had risen to almost 200, but his confidence level was higher. He had subsequently started a simple system for notifying people that a project had ended.

The TQM Transition

We have seen in Chaps. 4 through 9 how to develop the total quality implementation, as well as some hypothetical problems which arose in our example. As a project, the total quality implementation has a defined end point. However, to achieve quality in all aspects of a business requires constant attention, evaluation, and periodic goal review. The implementation has ended, but there must be continuous improvement.

We are now at the transition point, where several activities should take place to "close out" the project. Quality becomes the rule rather than the exception. As stated earlier in this chapter, there are a number of items to be considered:

1. Evaluate the objectives.
2. Record lessons learned.
3. Close vendor contracts.
4. Turn off the money.

5. Hand off to users.
6. Thank the project team.

We can take each of the six items and see how it applies to our implementation example.

Evaluating the objectives

Two objectives were identified in Chap. 3 as a result of the total quality assessment. First, there needed to be a reduction in variation in two key manufacturing processes. The associated deliverable and milestone was the data showing variation reduction and 12/31/93, respectively. Second, there needed to be an improvement in communication to all employees. The deliverable for this objective was a communication survey showing 90 percent or greater satisfaction, and the milestone was 3/31/93.

There needs to be a formal review of the project objectives by both executive management and the project team. In this case, the president, C. P. Kaye, the QA manager, and the HR manager will all be involved.

Top management must be satisfied that the objectives have been accomplished in terms of schedule, cost, and performance. What happens if any combination of these three categories has not been achieved? Will top management hit the roof and begin executing those responsible? If our project owner and leaders have truly managed the project, schedule, cost, and performance, issues will not be a surprise to top management. In fact, top management will have authorized continuing the project in light of schedule and/or cost overruns. They will be deciding simply if the project is complete. Were the objectives accomplished? If there has been a reduction in variation and a 90 percent or greater satisfaction rating in communication, the project is complete.

Recording the lessons learned

Documenting what went right and what went wrong is important for future projects of this nature. A total quality implementation occurs only once, but similar phases, activities, and tasks will surface in the future. Part of the total quality concept includes improvement projects, and these will always exist. Having a written record provides a baseline from which to work for those involved with quality projects.

So, what lessons can be learned from our example project? The only problem which surfaced in this case was the mechanical breakdown at the printer. In evaluating the situation we may, in retrospect, see some options for avoiding the problem. First, there may be a need to evaluate printers more thoroughly during the contract period. A visit to the printer could possibly have revealed the potential for a breakdown. Second, selecting a backup printer to "wait in the wings" could also be an option. Third, at the time of the dilemma a solution was suggested to do the printing in-house. Future projects may be best served by scheduling this activity rather than hiring a printer.

Closing vendor contracts

There are three contracts in our total quality implementation which need to be closed. If services were contracted for the communications media, it will need to be closed. The printing contract will need to be closed, although this was probably a purchase order. Finally, the consulting services contract will most likely need to be closed. In this case, the contract may be extended to assist with other quality projects.

Turning off the money

Turning off the money requires a closing of all accounts related to the total quality project. There will no longer be money available for printing, training (as it relates to the TQM implementation), surveys, and consulting services. Red flags will appear if charges are made to any of the closed project accounts, alerting management to potential financial problems. Consultants are notorious for slipping between the accounting cracks, receiving checks months or years after a particular project has ended. (Just kidding, of course!)

Handing off to users

This may be considered one of the most important aspects of closing the total quality project. Similar to the handoff in a relay race, a great deal of gain can be lost with a poor or missed handoff. A system must be in place to assure the continued use of statistical methods and to periodically gauge the effectiveness of communications. Top management makes its biggest mistake by assuming that this is the end, that we now have quality. The total quality implementation is merely the launching pad for continuous improvement, not the final effort. Without tracking and follow-up, TQM is doomed to fail. The handoff and associated systems (if applied) will prevent "See, I told you it wouldn't work."

Thanking the project team

Another important aspect of the total quality transition to use is acknowledging, recognizing, and rewarding people for the work they have performed. This holds true for any type of work, but it is especially important in projects. Projects are a unique, one-time effort in which people are usually given an extra hat to wear. The total quality project has created additional stress to all involved. The gratuity can take a number of forms. It might be a banquet or several small gatherings for the individual project teams. Bonuses are sometimes given, but money alone is not generally accepted as a good way to show appreciation. Plaques or other tokens may be given at a formal ceremony. Whatever the decision, it should fit with the organizational culture of the company.

Epilogue

Implementing total quality by using the process of project management will always follow the same path. The eight steps described in the book are required for complete implementation. If one or more of the steps is bypassed, then the project is most certainly placed on the endangered species list. Describing the project establishes the objectives and a baseline from which to work. Assembling the project team is essential to proj-ect success, and careful selection of the members is paramount. The work breakdown structure then provides all the packages of work to be performed throughout the project. Estimating task durations along with resource allocation and associated costs broadens the plan and provides more detail for the project. The schedule can then be calculated with a logic network to provide a time road map. Starting the project means the beginning of plan execution, where resources are given the authorization to proceed. As the project continues, tracking progress, reporting problems, and adjusting the plan through the use of several tools is accomplished. Finally, a complete review of the project, including the good and bad aspects of the project, is conducted and documented. Quality does not end here, however. It continues through persistence and further effort.

Total quality and project management initially seem to be unrelated disciplines. However, we have seen throughout the book how the process of project management augments the effectiveness of the total quality implementation. Taking the step-by-step approach eliminates some of the guesswork and creates a system from which to work. Create a plan, execute it, and make changes when necessary. It seems relatively simple, but consider how often failures occur. Sometimes a plan is never even considered. To charge ahead and just do it may be acceptable in certain circumstances, but not with TQM. Sometimes a plan is developed and never followed, which means the plan was a waste of time in the first place. There must be a commitment to quality in order for it to work and continue to work.

Index